ANALYSIS AND DESIGN OF SWIRL-AUGMENTED HEAT EXCHANGERS

ANALYSIS AND DESIGN OF SWIRL-AUGMENTED HEAT EXCHANGERS

Editor
V. M. Ievlev
USSR Academy of Sciences

English Edition Editor
T. F. Irvine
SUNY, Stony Brook

HEMISPHERE PUBLISHING CORPORATION
A member of the Taylor & Francis Group
New York Washington Philadelphia London

Co-authors: **Yu. N. Danilov, B. V. Dzyubenko, G. A. Dreitser, and L. A. Ashmantas**

ANALYSIS AND DESIGN OF SWIRL-AUGMENTED HEAT EXCHANGES

Originally published as Teploobmen i gidrodinamika v kanalakh slozhnoy formy by Mashinostroyeniye, Moscow, 1986.

Translated by E. A. Zharkova.

1 2 3 4 5 6 7 8 9 0   E B E B   9 8 7 6 5 4 3 2 1 0

This book was set by Desktop Publishing.
Cover design by Sharon Martin DePass.
Edwards Brothers, Inc. was the printer and binder.

Library of Congress Cataloging-in-Publication Data

Teploobmen i gidrodinamika v kanalakh slozhnoĭ formy. English.
Analysis and design of swirl-augmented heat exchangers / editor V. M. Ievlev, English edition editor T. F. Irvine ; [co-authors, Yu. N. [i.e. Yu. I.] Danilov . . . et al. ; translated by E. A. Zharkova].
p.   cm.
Translation of: Teploobmen i gidrodinamika v kanalakh slozhnoĭ formy.
Includes bibliographical references.

1. Heat—Transmission. 2. Hydrodynamics. 3. Heat exchangers—Design and construction. I. Ievlev, V. M. (Vitaliĭ Mikhaĭlovich), 1926-   . II. Irvine, Thomas F. (Thomas Francis) III. Danilov, IU. I. IV. Zharkova, E. A. V. Title.
TJ260.T36213   1990
621.402′5—dc20   89-19863
CIP

ISBN 0-89116-701-3

CONTENTS

PREFACE

Heat exchangers nowadays are widely used in aviation engineering for cooling aviation engine systems, vehicle members, aircraft instrument compartments, and cabins [30].

Such heat exchangers must be small in size and mass, have minimum hydraulic losses, and be highly reliable in operation. These demands also hold true for heat exchangers employed in power engineering, chemical, and other branches of industry, since these account for the essential part of power plants and basic equipment by mass and volume. Thus, in chemical and petroleum refinery industries, heat exchangers account for a major proportion of the mass and cost of the basic plant.

Also, the share of heat exchanger mass and volume is appreciable in heat engines. Heat transfer augmentation in channels due to flow swirling [56, 57] is among one of the more promising ways to design compact heat exchangers. In [7], intricate heating surfaces shaped as helical ovals and three-petal tubes were proposed to increase the efficiency and reliability of power-stressed installations. A crossflow helical tube heat exchanger was considered in [6] where it was shown to be possible to improve heat transfer at small twisting pitches of tube blades for moderate pressure losses.

Enhancement of convective heat transfer and flow mixing is, at present, one of the urgent problems. Its solution is of great scientific and practical importance. Decreasing the size and mass of heat exchangers, the amount of metal spent for their production, and their cost improves crossflow mixing, reduces cross-sectional temperature nonuniformity, and diminishes heat exchanger surface heating.

Heat transfer in circular tube bundles can be mainly enhanced by increasing a heat carrier velocity, which is far from being beneficial because of drastic energy consumption. The development of artificial roughness, the use of corrugated channels, and the mounting of the wire, diaphragms, washers, and rods in the channels hinder heat transfer increase as opposed to hydraulic resistance and energy consumption for heat carrier pumping.

In helical oval tube heat exchangers [2, 13, 24, 58, 59], convective heat transfer is enhanced by flow swirling in channels intricately shaped by a close-packed bundle of such tubes. In this case, success is attained not only in improving heat transfer due to flow swirling both inside the helical tubes and in the intertube space, but also in a substantial increase of heating surfaces per unit volume of the apparatus. Higher thermal and hydraulic characteristics can be obtained in helical tube heat exchangers, as compared to circular tube apparatus, since helical heat carrier swirling in complex-geometry channels originates transverse velocity components, additional turbulization, and secondary circulation of the flow. These mechanisms cause an intensive exchange of liquid portions between the wall layer and the flow core, thus improving heat and mass transfer.

The spiral cross flow in the intertube space of a helical tube heat exchanger allows a temperature field to be levelled in the intertube space and at its outlet; and to increase apparatus efficiency and its operational reliability. Moreover, the spiral flow swirling substantially reduces wall temperature nonuniformity over the perimeter of helical tubes.

This monograph proposes the physically grounded helical tube flow models and their mathematical description. Some methods of approximate closure of the equation systems for the flow in the core and in the wall layer are considered. The possibility of using the methods of calculating a boundary layer at the entrance length of a bundle is shown. The similarity and dimension theories were adopted to propose a new similarity number to take into account specific features of the flow in a helical tube bundle and to correlate the experimental data on heat transfer, interchannel heat carrier mixing, hydraulic resistance, and flow structure. The same similarity number was obtained using the semiempirical turbulence theory.

The proposed flow model based on the effective wall layer thickness allows the laws for heat transfer and friction in the circular tubes to be used to generalize experimental data on heat transfer and hydraulic resistance in helical tube bundles. Such experimental data processing offers ample scope for modelling and diminishes the number of experiments to establish criterial relations. The flow model for a homogenized medium replacing a real tube bundle is experimentally substantiated. The methods are proposed to solve the systems of flow equations for axisymmetric and asymmetric heat supply nonuniformity. The criterial relations are specified to calculate the effective turbulent viscosity coefficient and thermal conductivity that enter these equations. Experimental methods to study heat transfer and flow dynamics are developed.

The discovered specific features of the flow made it possible to substantiate thermal and hydraulic characteristics depending on basic similarity numbers. The criterial relations for heat transfer and hydraulic resistance in longitudinal and crossflows past and inside helical tube bundles are employed to estimate the efficiency of heat exchangers with such tubes. The heat exchanger efficiency estimated by the elaborated methods has shown that, at the assigned heat power and the same hydraulic losses, the helical tube bundles, when used instead of the straight circular ones, offer about a 20 to 30% decrease of the heat exchanger mass and volume.

Preface, Chapters 1, 3, and 4 were written by B. V. Dzyubenko, Chapters 2 and 5 were written by B. V. Dzyubenko, G. A. Dreitser, and Yu. I. Danilov, Chapters 6 and 7 were written by L. A. Ashmantas, and Chapter 8 was written by G. A. Dreitser.

NOMENCLATURE

a	thermal diffusivity, coefficient for the flow structure
b	"half" width of a jet, $b = 2r_{\text{mean}}$
c_p	heat capacity
d	maximum size of a tube profile
d_{eq}	equivalent diameter
D_t	effective heat diffusion coefficient
E	wire voltage drop
e'	voltage fluctuation
F_{f}	area of the flow cross section of a tube bundle
G	air mass flowrate
G_i	axial heat carrier flowrate in a cell
G_{ij}	heat carrier flow in the transverse direction from a cell i to a cell j per unit length of a channel
I	enthalpy of unit mass
j_0	enthalpy head in a boundary layer
k	dimensionless effective diffusion coefficient
$\bar{k}$	mean value of the coefficient k
L_{E}, L_{L}	spatial integral turbulence scales for the flow according to Euler and Lagrange
l_{c}	control section length
l	mixing path
m	bundle porosity with respect to heat carrier

q	energy flow in a boundary layer; heat flux density
q_v	volumetric density of heat release
p	pressure
r	radial coordinate
r_k	bundle radius
r_{mean}	"half" jet radius
s	tube twisting pitch
T	temperature
t	time
t_{mean}	mean temperature head in a boundary layer
u, v, w	averaged velocity components in the orthogonal coordinate system
u', v', w'	pulsational velocity components
u_t, u_r	tangential and radial velocity components in cylindrical coordinates
V	averaged velocity vector modulus
v_1	mean quadratic pulsational velocity
v_c	velocity vector component along the x_c axis (equations are written in a tensor form)
$w^2 = \overline{(v_i' v_i')}/3$	pulsational squared velocity to a turbulence isotropy approximation
x_k (*where* $k = 1, 2, 3$)	Cartesian coordinates
$\bar{y}^2$	mean statistical squared displacement
α	heat transfer coefficient, dimensionless friction coefficient
α_m	dimensionless heat transfer coefficient
Γ	flow swirling degree
Δp	pressure drop
δ	wall layer thickness
δ^*	displacement thickness of a boundary layer
ϵ	effective tubulence intensity
λ	thermal conductivity
μ	dynamic viscosity coefficient, coefficient for interchannel flow mixing (Chapter 4)
η	second viscosity coefficient
ν	kinematic viscosity coefficient
ν_{eff}	effective turbulent viscosity coefficient
λ_{eff}	effective thermal conductivity
ξ	hydraulic resistance coefficient
ρ	density
τ	shear stress
τ_{xw}	axial component of wall shear stress
τ_{zw}	tangential component of wall shear stress
τ_{z0}	tangential component of shear stress along the channel axis
$\tau_{\Sigma w}$	total wall shear stress

φ	angular coordinate, angle between the anemometer wire and the velocity vector direction
φ'	fluctuation of an angle φ
ψ	temperature factor (Chapters 6 and 7), bundle porosity with respect to heat carrier (Chapter 8)
Fr_M	number for the specific features of the flow in a helical tube bundle
Nu	Nusselt number
Pe	Peclet number
Pr	Prandtl number
Re	Reynolds number

SUBSCRIPTS

in	inside a tube
hot	hot side of a heat exchanger
max	maximum, modified
c	control, casing
out	outside a tube
o	on the bundle axis
f	flow
w	wall
mean	mean mass
t	turbulent
tube	tube
cold	cold side of a heat exchanger
c	centrifugal
1, 2	experimental section inlet and outlet
I, II	heat exchanger inlet and outlet, respectively
d	determined with respect to d_{eq}
m	at a mean wall layer temperature
δ	determined with respect to a wall layer thickness

CHAPTER
ONE

SPECIFIC FEATURES OF THE PROCESSES OF HEAT TRANSFER AND FLOW IN COMPLEX-GEOMETRY CHANNELS

1.1 SPECIFIC FEATURES OF THE DESIGN OF A HELICAL TUBE HEAT EXCHANGER

The specific features of heat transfer and fluid dynamics in complex-geometry channels formed by helical tube bundles are defined by the structural characteristics of these bundles. Figure 1.1 is a schematic of a shell-tube heat exchanger with oval-shaped helical tubes, with their straight round ends fastened into the tube plates*. The tubes in this heat exchanger are located relative to each other such that they have contact over a maximum amount of the oval. A spiral swirling of the fluid occurs by its circulation in the tubes and in the intertube spaces. The flow in the intertube space is of a more complex nature, which may be conventionally considered as a system of alternating interconnected spiral and through channels. The turbulence in such a system is generated by the fixed wall and is due to the friction between liquid layers with different velocities. The flow is swirled in opposite directions in the spiral channels of adjacent tubes, which results in discontinuities of the tangential velocity components. The longitudinal velocity components in the flow core also undergo tangential discontinuities due to different flow conditions in the through channels and behind the locations of contacting adjacent tubes. The flow in a helical tube bundle is also affected by secondary circulations due to the centrifugal forces that exist with fluids flowing in spiral channels. Thus, the turbulence in a helical tube bundle enhances heat transfer and interchannel flow mixing.

---

*B. V. Dzyubenko and Yu. V. Vilemas. A shell-tube heat exchanger. Author's Certificate No. 761820 (USSR). Bulletin of Inventions, 1980, No. 33, p. 194.

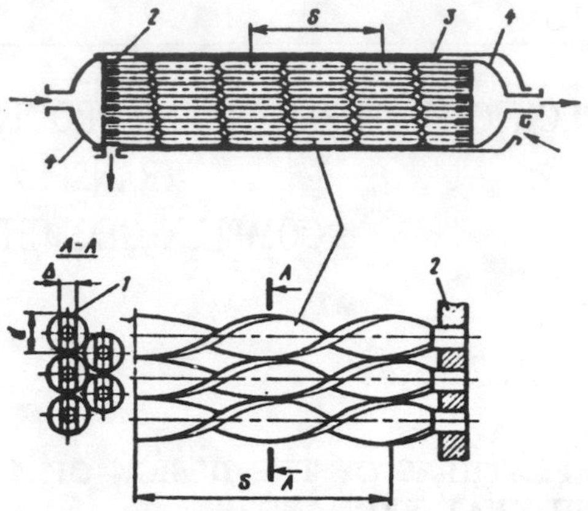

Figure 1.1 Swirled flow heat exchanger: *1*) helical tubes; *2*) tube plates; 3) tube shell; 4) bottom.

It is a characteristic of this type of flow that flow swirling is produced along the entire tube bundle and therefore the flow structure is stabilized at some distance from the inlet. This results in a stabilization, on the average, of the coefficients of heat transfer, hydraulic resistance, and mixing, although the heat transfer coefficient may vary along the perimeter of a complex channel, such as a helical tube bundle. Within some limits, the heat transfer coefficient may also periodically change along the tube bundle because of the periodic spacing of the tubes in contact.

The flow in a helical tube bundle is spatial, i.e., together with a longitudinal velocity vector component there exist transverse velocity components that greatly enhance interchannel flow mixing in the bundle. The high turbulence level of the flow, convective transfer in a cell, and ordered transfer in the cross section of the bundle due to spiral flow swirling by means of twisted tubes are the mechanisms responsible for the specific features of cross flow mixing in a bundle, as compared to the transport phenomena in a straight circular tube.

The most important geoemtrical parameter of a helical tube bundle is the blade twisting pitch *s* based on the maximum oval-shaped tube size *d*. This parameter, to a considerable degree, specifies the intensity of the centrifugal force field in the bundle as well as the characteristic properties of heat transfer and mixing of the heat transfer fluid. An optimum relative twisting pitch s/d is determined only when maximum improvement of these processes is provided at acceptable values of the hydraulic resistance coefficients for a bundle. In this case, it is possible to design rather compact helical tube heat exchangers. Solving the urgent problems of convective heat transfer and mixing enhancement due to circular tubes being replaced by oval-shaped ones enables one not only to substantially decrease the overall dimensions and mass of heat exchangers, their metal consumption, and cost, but also to reduce temperature nonuniformities over the heat exchanger cross section and to depress the heating of a heat exchanger surface. Such heat exchangers can be successfully used in aviation engineering as well as in different branches of industry.

The analyzed details of helical tube bundles and the specific features of the flow in these bundles may also give rise to a particular behavior of heat transfer and hydrodynamic processes as a function of the Reynolds number. In fact, from general considerations, the spiral swirling of the fluid in the transition range of Reynolds numbers must, to a great extent, improve the heat transfer and mixing at high Reynolds numbers. The mechanisms of these processes must be experimentally specified.

The swirling spiral flow initiates laminar flow instabilities at Reynolds numbers which are approximately, by an order and a half, less than those in a straight circular tube. However, unlike the flow in coils, the swirling flow in the intertube space of a helical tube heat exchanger also results in an earlier transition to turbulent flow (at $Re \approx 10^3$). In this case, the transition from laminar to turbulent flow in helical tube bundles is smooth.

In the present investigation, fully developed flow models were adopted to generalize the comprehensive data on the flow structure, heat transfer, hydraulic resistance, and interchannel fluid mixing. Also, experimentally based methods were proposed for the thermal and hydraulic calculations of helical tube heat exchangers with regard to cross mixing. In this case, preference was given to the longitudinal flow past close-packed helical tube bundles, although much attention was also paid to other structural schematics of helical tube heat exchangers. Figure 1.2 shows a schematic of a shell-tube apparatus* which differs from the previous one. In this device, a helical tube bundle is twisted relative to its longitudinal axis so that the relative twisting pitch decreases with respect to

*B. V. Dzyubenko, Yu. V. Vilemas, R. R. Varshkyavičius and G. A. Dreytster. A shell-tube heat exchanger. Author's Certificate No. 937954 (USSR). Bulletin of Inventions, 1982, No. 23, p.189.

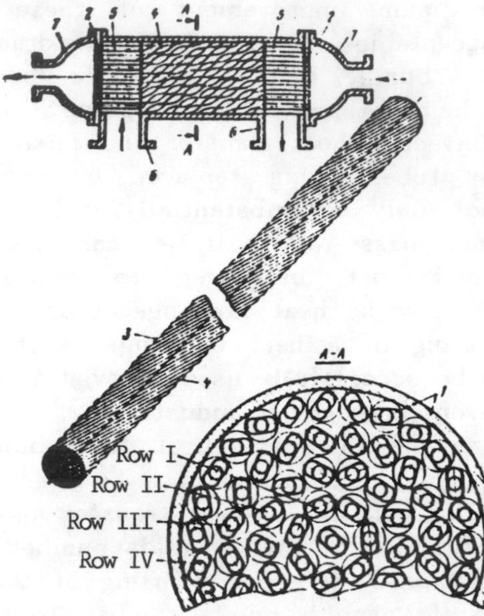

Figure 1.2 Twisted helical tube bundle heat exchanger: *1*) tube shell; *2*) tube plate; *3*) twisted tube bundle; *4*) helical tube; *5*) straight round tube ends; *6, 7*) connections.

the bundle radius, and the lengths of the straight tube ends are equal to the diameter of the inlet and outlet connections. The connecting channels are formed in the tube shell. The porosity of these channels is higher than that of the twisted part of the bundle. In this case, the heat transfer and mixing of the fluid are improved, and the inlet velocity and temperature fields are smoothed out by the lateral supply and removal of the heat transfer fluid. In addition, a relative tube deformation may be expected with varying temperatures. The helical tube twisting relative to the bundle axis initiates a radius redistribution of the fluid flow in the intertube space because of the difference in hydraulic resistance coefficients for the tube rows located on different radii. Such a heat exchanger design permits suppression of azimuthal nonuniformities of velocity and temperature that develop, in particular, in the lateral fluid flow.

Figure 1.3 shows a schematic of a crossflow helical tube heat exchanger. The design of this heat exchanger should differ substantially from a crossflow circular tube heat exchanger

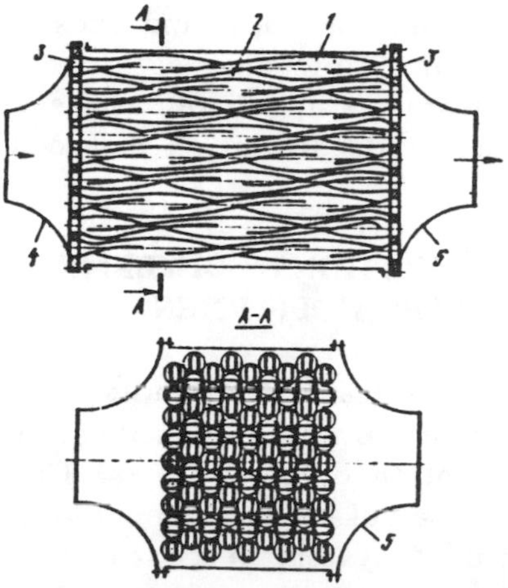

Figure 1.3 Crossflow helical tube heat exchanger: *1)* helical tube; *2)* slotted channels between tubes; *3)* tube plate; *4, 5)* connecting channels for supply and removal of heat carriers (fluids).

with regard to enhancement of heat transfer and leveling temperature nonuniformities over the tube perimeter. First, this heat exchanger contains a bundle of helically twisted oval-profile tubes in contact with adjacent ones over the maximum size of the oval, i.e., a close-packed bundle, which cannot be implemented in the case of circular tubes. Moreover, the helical tubes are specially arranged in a bundle, namely: the tubes in each transverse row of a bundle are mounted with spacers forming, along the tube bundle, slotted channels with a maximum width equal to half the difference between the maximum and minimum sizes of the oval and only contact tubes of adjacent rows*. Such a tube arrangement affords optimum augmentation of heat transfer from the tubes in crossflow due to spiral fluid swirling along the helical tubes, i.e., normal to the main flow. In this case, the tempeature field in the intertube space and at its outlet is levelled, which improves the operational reliability of the heat exchanger. As shown in Chapter 7, in a heat exchanger of such a design, with a relative helical tube twisting pitch s/d=6-12, the temperature nonuniformity along the perimeter is 15-20%, respectively, which is 2-3 times smaller than for circular tube bundles.

*B. V. Dzyubenko, G. A. Dreytser, Yu. V. Vilemas, N. N. Paramonov, L. A. Ashmantas, and Yu. V. Survila. A shell-tube heat exchanger. Author's Certificate No. 840662 (USSR), 1981, No. 23, p. 178.

The above specific features of heat exchangers with differently arranged helical tubes, and those features associated with heat transfer and flow processes are taken into account to make experimental studies of these processes and to develop flow models and calculation methods.

1.2 GOVERNING EQUATIONS. DETERMINATION OF TURBULENT TRANSFER COEFFICIENTS

The fluid flow in a helical tube bundle is spatial in nature. A general system of differential equations may be used for a mathematical description of such flows. The macroscale of the flow in a bundle is equal to the diameter of the tube shell in which it is located. A system of continuity, Navier-Stokes, and energy equations, and ρ, μ, λ, c_p as a function of T and p must be solved to determine the three components of velocity (u,v,w), pressure (p), temperature (T), and the parameters for the physical properties $(\mu, \lambda, c_p, \rho)$ of a fluid at any time instant. The above equations are valid both in the laminar and turbulent regimes at instantaneous values of these parameters.

A turbulent swirled flow in geometrically complex channels formed by helical tube bundles, as well as the flow inside these tubes, may, in general, be described by the equations of motion, energy, and continuity which in a uniform medium with variable physical properties assume the following tensor form [23]:

$$\rho \frac{\partial v_i}{\partial t} + \rho v_k \frac{\partial v_i}{\partial x_k} = -\frac{\partial p}{\partial x_i} + \frac{\partial}{\partial x_k}\left\{\mu\left(\frac{\partial v_i}{\partial x_k} + \frac{\partial v_k}{\partial x_i}\right)\right\} + \frac{\partial}{\partial x_i}\left\{\left(\eta - \frac{2}{3}\mu\right)\frac{\partial v_e}{\partial x_e}\right\}, \quad i = 1, 2, 3 \tag{1.1}$$

$$\rho \frac{\partial I_0}{\partial t} + \rho v_k \frac{\partial I_0}{\partial x_k} = \frac{\partial p}{\partial t} + \frac{\partial}{\partial x_k}\left\{\mu v_i \left(\frac{\partial v_i}{\partial x_k} + \frac{\partial v_k}{\partial x_i}\right) + \left(\eta - \frac{2}{3}\mu\right) v_k \frac{\partial v_l}{\partial x_l} + \lambda \frac{\partial T}{\partial x_k}\right\} \tag{1.2}$$

$$\frac{\partial \rho}{\partial t} + \frac{\partial (\rho v_k)}{\partial x_k} = 0 \tag{1.3}$$

where $$I_0 = I + \frac{v^2}{2} \tag{1.4}$$

v is the total velocity vector modulus

$$\rho,\ I,\ \mu,\ \eta,\ \lambda = f(p, T) \tag{1.5}$$

In Eqs. (1.1)-(1.3), a summation is proposed with respect to the iterated coordinate indices along all the coordinate axes, for example,

$$v_k \frac{\partial I_0}{\partial x_k} = \sum_{k=1}^{3} v_k \frac{\partial I_0}{\partial x_k} = u\frac{\partial I_0}{\partial x} + v\frac{\partial I_0}{\partial y} + w\frac{\partial I_0}{\partial z}$$

and $$\frac{\partial v_l}{\partial x_l} = \operatorname{div} \mathbf{V} = \frac{\partial u}{\partial x} + \frac{\partial v}{\partial y} + \frac{\partial w}{\partial z}$$

In addition to the above equations, the geometrical conditions specifying the shape and sizes of a helical tube bundle and the boundary and initial conditions (physical, boundary-value, and time) must be specified.

To make a mathematical study of turbulent flow, it is resolved into averaged and pulsational motions: $u=\bar{u}+u'$, $v=\bar{v}+v'$, $w=\bar{w}+w'$, $p=\bar{p}+p'$, $T=\bar{T}+T'$, i.e., the flow is represented as a separate turbulent mode superimposed on the averaged fluid motion. Here, the bar above u, v, and the other variables means that the variable is time-averaged.

The continuity, Navier-Stokes, and energy equations for turbulent flow must be satisfied not only at any moment of time, but also on the average. In this case, the time-averaged values and those at a fixed point of the space are implied, i.e.,

$$\bar{u} = \frac{1}{t}\int_{t_0}^{t_0+t} u\,dt \quad \text{etc.}$$

the time interval t is so large that $\bar{w}'=0$, $\bar{v}'=0$, $\bar{w}'=0$, $\bar{p}'=0$, etc.

The effect of the pulsating motion on the averaged one manifests itself in increasing the deformation resistance of the averaged motion. The continuity, motion, and energy equations, which are satisfied by the time-averaged velocities $\bar{u}$, $\bar{v}$, $\bar{w}$, pressure $\bar{p}$, and temperature $\bar{T}$, are derived from a system of the continuity, Navier-Stokes, and energy equations using relevant averaging rules [23, 54].

Since the present book deals only with steady-state flows in tube bundles, consideration will be restricted, on the whole, to the steady-state flow of a medium. Thus, upon substituting instantaneous values of velocity, enthalpy, and other quantities with the sum of their mean and pulsating values, Eqs. (1.1)-(1.4) reduce to the system for the averaged flow of a medium [23]:

$$\overline{\rho v_k}\frac{\overline{\partial v_l}}{\partial x_k} = -\frac{\partial \bar{p}}{\partial x_l} + \frac{\partial}{\partial x_l}\overline{\left(\eta - \frac{2}{3}\mu\right)\frac{\partial v_l}{\partial x_l}} + \frac{\partial}{\partial x_k}\left\{\bar{\mu}\left(\frac{\partial \bar{v}_l}{\partial x_k} + \frac{\partial \bar{v}_k}{\partial x_l}\right)\right.$$
$$\left. + \overline{\mu'\left(\frac{\partial \bar{v}_l'}{\partial x_k} + \frac{\partial \bar{v}_k'}{\partial x_l}\right)} - \overline{(\rho v_k)' v_l'}\right\} \tag{1.6}$$

$$\overline{\rho v_k}\frac{\partial I_0}{\partial x_k} = \frac{\partial}{\partial x_k}\left\{\overline{\bar{\mu} v_l\left(\frac{\partial v_l}{\partial x_k} + \frac{\partial v_k}{\partial x_l}\right)} + \overline{\mu'\left[v_l\left(\frac{\partial v_l}{\partial x_k} + \frac{\partial v_k}{\partial x_l}\right)\right]'}\right.$$
$$\left. + \overline{\left(\eta - \frac{2}{3}\mu\right) v_k \frac{\partial v_e}{\partial x_e}} + \frac{\bar{\lambda}}{c_p}\frac{\partial \bar{I}}{\partial x_k} + \overline{\left(\frac{\lambda}{c_p}\right)'\frac{\partial I'}{\partial x_k}} - \overline{(\rho v_k)' I_0'}\right\} \tag{1.7}$$

$$\frac{\partial \overline{(\rho v_k)}}{\partial x_k} = 0 \tag{1.8}$$

$$\bar{I}_0 = \bar{I} + \frac{\bar{v}^2}{2} + \frac{\overline{v'^2}}{2} \tag{1.9}$$

The continuity equation (1.8), which is satisfied by the averaged velocity components (assuming that the correlation between ρ' and u' is zero), is of the following form:

$$\frac{\partial(\rho u)}{\partial x} + \frac{\partial(\rho v)}{\partial y} + \frac{\partial(\rho w)}{\partial z} = 0$$

and the equations of motion (1.6) assume the form which was first obtained by Reynolds:

$$\rho u\frac{\partial u}{\partial x} + \rho v\frac{\partial u}{\partial y} + \rho w\frac{\partial u}{\partial z} = -\frac{\partial p}{\partial x} + \left(\frac{\partial \sigma_x}{\partial x} + \frac{\partial \tau_{xy}}{\partial y} + \frac{\partial \tau_{xz}}{\partial z}\right)$$
$$-\left[\frac{\partial(\rho\overline{u'^2})}{\partial x} + \frac{\partial(\rho\overline{u'v'})}{\partial y} + \frac{\partial(\rho\overline{u'w'})}{\partial z}\right]$$
$$\rho u\frac{\partial v}{\partial x} + \rho v\frac{\partial v}{\partial y} + \rho w\frac{\partial v}{\partial z} = -\frac{\partial p}{\partial y} + \left(\frac{\partial \tau_{xy}}{\partial x} + \frac{\partial \sigma_y}{\partial y} + \frac{\partial \tau_{yz}}{\partial z}\right)$$
$$-\left[\frac{\partial(\rho\overline{u'v'})}{\partial x} + \frac{\partial(\rho\overline{v'^2})}{\partial y} - \frac{\partial(\rho\overline{v'w'})}{\partial z}\right]$$
$$\rho u\frac{\partial w}{\partial x} + \rho v\frac{\partial w}{\partial y} + \rho w\frac{\partial w}{\partial z} = -\frac{\partial p}{\partial z} + \left(\frac{\partial \tau_{xz}}{\partial x} + \frac{\partial \tau_{yz}}{\partial y} + \frac{\partial \sigma_z}{\partial z}\right)$$

$$-\left[\frac{\partial(\rho\overline{u'w'})}{\partial x}+\frac{\partial(\rho\overline{v'w'})}{\partial y}+\frac{\partial(\rho\overline{w'^2})}{\partial z}\right]$$

In the energy equation, the effect of the pulsations is that the term for the heat flux due to molecular conductance per unit time per unit area normal to each coordinate axis should be supplemented with terms for the turbulent heat transfer. If the terms for friction and expansion (compression) work with varying volume are neglected, then for substantially subsonic flow the energy equation is obtained in the following form:

$$\rho uc_p\frac{\partial T}{\partial x}+\rho vc_p\frac{\partial T}{\partial y}+\rho wc_p\frac{\partial T}{\partial z}=\frac{\partial q_x}{\partial x}+\frac{\partial q_y}{\partial y}+\frac{\partial q_z}{\partial z}$$

where $q_x=\rho c_p\overline{u'T'}+\lambda\frac{\partial T}{\partial x}$, $q_y=\rho c_p\overline{v'T'}+\lambda\frac{\partial T}{\partial y}$, $q_z=\rho c_p\overline{w'T'}+\lambda\frac{\partial T}{\partial z}$

Since mathematical difficulties make it impossible to obtain a general solution to this system of equations for the flow in a helical tube bundle, the system must be simplified with regard to certain physical flow models. Below, different flow models will be analyzed.

For a mathematical description of the flow near helical tubes where viscosity forces manifest themselves, the assumptions common to boundary layer theory may be made. Let the wall layer variable along the tube perimeter be replaced by the one constant along a thickness δ and let the tube be shaped as a twisted plate. Since the wall layer thickness δ is small compared to the wall curvature radius of the spiral channel of a tube, and the curvature of the spiral surface of a tube is small and does not undergo sharp changes, the wall layer pressure p may be assumed constant with respect to δ. The flow in this case may be mathematically described using the equations for a flat boundary layer. When the boundary layer develops on a helical tube, its thickness varies from zero at the inlet to δ at the end of the entrance length and remains constant downstream. It is assumed that the flow conditions on the ends of the tube blades do not affect the one in the wall layer, because the area of the end surfaces is small in comparison with that of the spiral channels, and the vortices forming in the flow around the tubes in contact are entrained into the flow core and do not destroy the boundary layer on the spiral surface of the tubes.

Then, Eqs. (1.6)-(1.9) for boundary layer flow may be reduced to:

$$\rho u \frac{\partial u}{\partial x} + \rho v \frac{\partial u}{\partial y} = -\frac{dp}{dx} + \frac{\partial \tau}{\partial y} \tag{1.10}$$

$$\rho u \frac{\partial I_0}{\partial x} + \rho v \frac{\partial I_0}{\partial y} = \frac{\partial q_0}{\partial y} \tag{1.11}$$

$$\frac{\partial(\rho u)}{\partial x} + \frac{\partial(\rho v)}{\partial y} = 0 \tag{1.12}$$

In (1.10)-(1.12) and below, the averaging signs of the parameters, $\bar{u}$, $\bar{v}$, etc. are omitted and

$$I_0 = I + \frac{u^2}{2} \tag{1.13}$$

$$\tau = \rho \nu \frac{\partial u}{\partial y} - \rho \overline{u'v'} \tag{1.14}$$

$$q_0 = q + u\tau \tag{1.15}$$

$$q = \rho a \frac{\partial I}{\partial y} - \rho \overline{v'I'}$$

τ is the total shear stress and q is the heat flux density (with no regard to q_{rad}). The direction of the wall heat flux q is positive. This system of equations must be supplemented with the relation ρ, ν, a = φ (p, I). In Eq. (1.9), neglecting the quantity $\overline{v'^2}/2$, it is assumed that $v^2 \approx v_1^2 \approx u^2$. To close this system of equations, the turbulent shear stress and turbulent heat flux density are represented in the form:

$$-\rho \overline{u'v'} = \rho \nu_T \frac{\partial u}{\partial y}, \quad -\rho \overline{(v'I')} = \rho a_T \frac{\partial I}{\partial y}$$

In this case,

$$\tau = \rho(\nu + \nu_T) \frac{\partial u}{\partial y} \tag{1.16}$$

$$q = \rho(a + a_T) \frac{\partial I}{\partial y} = \rho \left(\frac{\nu}{\mathrm{Pr}} + \frac{\nu_T}{\mathrm{Pr}_T} \right) \frac{\partial I}{\partial y} \tag{1.17}$$

Such a representation is valid for equilibrium turbulence when the production and dissipation of turbulent energy are equal, as well as for similarity flows, which, as shown below, is satisfied to a first approximation by the heat transfer fluid flowing in the intertube space of a helical tube heat exchanger. Usually, the quantities ν_T and a_T are determined empirically, assuming that the turbulence characteristics at each flow point depend only on its local parameters at this point. In this case, these characteristics conform with those for the same local flow parameters of an incompressible liquid, i.e.,

$$\nu_T = \nu_T\left(\rho,\ \nu,\ \frac{\partial u}{\partial y},\ l\right),\quad a_T = a_T\left(\rho,\ \nu,\ \frac{\partial u}{\partial y},\ l,\ a\right)$$

Then

$$\frac{\nu_T}{\nu} = \varphi\left(\frac{l\sqrt{\tau/\rho}}{\nu}\right);\ \mathrm{Pr}_T = \mathrm{Pr}_T\left(\frac{l\sqrt{\tau/\rho}}{\nu},\ \mathrm{Pr}\right);\ \frac{l}{\delta} = f\left(\frac{y}{\delta}\right) \tag{1.18}$$

l/δ may be determined using the empirical Nikuradse formula:

$$\frac{l}{\delta} = 0.14 - 0.08\left(1 - \frac{y}{\delta}\right)^2 - 0.06\left(1 - \frac{y}{\delta}\right)^4 \tag{1.19}$$

and ν_t/ν using the Reichardt formula:

$$\frac{\nu_T}{\nu} = 0.4\left\{\eta - 7.15\left[\mathrm{th}\left(\frac{\eta}{7.15}\right) + \frac{1}{3}\mathrm{th}^3\left(\frac{\eta}{7.15}\right)\right]\right\} \tag{1.20}$$

Equation (1.20), derived for an incompressible fluid, may be adopted for the flow of an arbitrary medium with variable properties if instead of $\eta = yv_*/\nu$, where $v_* = \sqrt{\tau/\rho}$, η is substituted in the form:

$$\eta = \frac{l\sqrt{\tau/\rho}}{0.4\nu} \tag{1.21}$$

assuming that near the wall $l \approx 0.4y$ and using the local values of τ, ρ, and ν. The quantity ν_T may also be found from the Prandtl formula $\nu_T = l^2\ (\partial u/\partial y)$, where the mixing length l is determined from Van Driest's expression:

$$\frac{l}{y} = 0.4\left[1 - \exp\left(-\frac{\eta}{26}\right)\right] \tag{1.22}$$

In this case, the calculating equation is obtained [23]

$$\frac{\nu_T}{\nu} = \frac{1}{2}\left\{\sqrt{1 + 0.64\eta^2[1 - \exp(-\eta/26)]^2} - 1\right\} \tag{1.23}$$

When the above equation is used, the calculation results differ only slightly from those obtained by Eq. (1.20).

Since in helical tube heat exchangers gases serve more frequently as the heat carrier, the turbulent Prandtl number

$$\mathrm{Pr}_{\mathrm{T}} \approx 1 \tag{1.24}$$

may be assumed. Thus, Eqs. (1.16), (1.17), (1.19), (1.20), (1.21), and (1.24) permit closing the system of equations for a turbulent boundary layer. It should be borne in mind that this approach to thermal and hydraulic calculations for a bundle should be considered as a first approximation.

Empirical methods and semiempirical expressions may also be adopted to determine ν_T and a_T. In [23], the turbulence balance equations for the intermediate region of a boundary layer, where the turbulent energy production is approximately equal to the dissipation, were used to obtain expressions for ν_T and a_T:

$$-\overline{u'v'}\frac{\partial u}{\partial y} \approx \frac{\nu}{2}\overline{\left[\left(\frac{\partial v_i'}{\partial x_\kappa}+\frac{\partial v_\kappa'}{\partial x_i}\right)\left(\frac{\partial v_i'}{\partial x_\kappa}+\frac{\partial v_\kappa'}{\partial x_i}\right)\right]} \tag{1.25}$$

$$-\overline{v'I'}\frac{\partial I}{\partial y} \approx a\overline{\left(\frac{\partial I'}{\partial x_\kappa}\frac{\partial I'}{\partial x_\kappa}\right)} \tag{1.26}$$

In this case, it is assumed that the quantities

$$+\overline{v'I'} \approx \sqrt{\overline{I'^2}}\cdot w, \qquad \overline{u'v'} \approx w^2$$

The right-hand side of (1.25) at low Reynolds numbers is

$$\overline{\left[\left(\frac{\partial v_i'}{\partial x_\kappa}+\frac{\partial v_\kappa'}{\partial x_l}\right)\left(\frac{\partial v_i'}{\partial x_\kappa}+\frac{\partial v_\kappa'}{\partial x_i}\right)\right]} \sim \frac{\nu w^2}{l^2}$$

where l is the size of large vortices at high Reynolds numbers:

$$\overline{\left[\left(\frac{\partial v_i'}{\partial x_\kappa}+\frac{\partial v_\kappa'}{\partial x_i}\right)\left(\frac{\partial v_i'}{\partial x_\kappa}+\frac{\partial v_\kappa'}{\partial x_i}\right)\right]} \sim \frac{\nu_M w^2}{l^2}$$

where ν_M is the effective kinematic viscosity coefficient that specifies the effect of small turbulent vortices on the large-scale pulsations.

At arbitrary Reynolds numbers it is assumed

$$\overline{\left[\left(\frac{\partial v_i'}{\partial x_\kappa}+\frac{\partial v_\kappa'}{\partial x_l}\right)\left(\frac{\partial v_i'}{\partial x_\kappa}+\frac{\partial v_\kappa'}{\partial x_l}\right)\right]} \sim (\nu+\nu_M)\frac{w^2}{l^2}$$

Similarly,

$$a\left[\overline{\frac{\partial I'}{\partial x_\kappa}\cdot\frac{\partial I'}{\partial x_\kappa}}\right]\sim(a+a_м)\frac{\overline{I'^2}}{l^2}$$

If one assumes

$$\nu_м=\alpha\nu_т,\ a_м=\beta a_т \tag{1.27}$$

where

$\alpha,\ \beta=\text{const}\ (\alpha,\ \beta<1)$

Substitution of (1.25) into (1.26) yields:

$$k_1 w^2\frac{\partial u}{\partial y}=A_1(\nu+\alpha\nu_т)\frac{w^2}{l^2} \tag{1.28}$$

$$k_2 w\sqrt{\overline{I'^2}}\,\frac{\partial I}{\partial y}=A_2(a+\beta a_т)\frac{\overline{I'^2}}{l^2} \tag{1.29}$$

where A_1, A_2, k_1, and k_2 are constants.

Assuming $\alpha A_1/k_1 = 1$ and substituting into (1.28) the quantity $\tau/\rho/(\nu+\nu_T)$ instead of du/dy from Eq. (1.28) we finally arrive at:

$$\frac{\nu_т}{\nu}=-\frac{1-\alpha}{2\alpha}+\sqrt{\left(\frac{1-\alpha}{2\alpha}\right)^2+\left(\frac{l\sqrt{\tau/\rho}}{\nu}\right)^2} \tag{1.30}$$

At $l\sqrt{\tau/\rho}/\nu >> 1$, from Eq. (1.30) we have $\nu_T/\nu \approx l\sqrt{\tau/\rho}/\nu$, which conforms with the Prandtl formula for turbulent friction [55]. Equation (1.30) yields the best agreement between theory and experiment at $\alpha = 0.15$. The expression for α_t is obtained in the following way. The quantity $\partial I/\partial y = k_2 w\sqrt{\overline{I'^2}}/a_T$ is substituted into (1.29). Then,

$$\frac{w^2}{a_т}=\frac{A_2}{k_2^2}(a+\beta a_т)\frac{1}{l^2} \tag{1.31}$$

A mean squared pulsational velocity w^2 may be expressed in terms of ν_T:

$$-\overline{u'v'}=k_1 w^2=\nu_т\frac{\partial u}{\partial y}$$

but $du/dy = \tau/\rho\ (\nu+\nu_T)$. Then,

$$w^2 = \frac{\nu_T \tau/\rho}{k_1 (\nu + \nu_T)} \tag{1.32}$$

From Eqs. (1.31), (1.32), and (1.29) at $\alpha A_1/k_1 = 1$, we have:

$$\frac{a_T}{\nu} = -\frac{1}{2\beta \mathrm{Pr}} + \sqrt{\frac{1}{4\beta^2 \mathrm{Pr}^2} + \frac{1}{\mathrm{Pr}_T^2} \frac{\nu_T}{\nu}\left(\frac{\nu_T}{\nu} + \frac{1}{\alpha}\right)} \tag{1.33}$$

where $\mathrm{Pr}_T = \sqrt{\beta k_1 A_2/k_2}$ is the turbulent Prandtl number far from the wall.

Instead of the above equation for a_T/ν, it is possible to use the relations [23]:

$$\mathrm{Pr}_T = \frac{\nu_T}{a_T} = \frac{\nu_T/\nu}{-\frac{1}{2\beta \mathrm{Pr}} + \sqrt{\left(\frac{1}{2\beta \mathrm{Pr}}\right)^2 + \frac{1}{\mathrm{Pr}_T^2} \frac{\nu_T}{\nu}\left(\frac{\nu_T}{\nu} + \frac{1}{\alpha}\right)}} \tag{1.34}$$

at $\nu_T/\nu \gg 1$, $\mathrm{Pr}_T \approx \overline{\mathrm{Pr}_T}$.

The above theory may be refined if turbulent diffusion due to molecular viscosity alone is allowed for. In this case, the "laminar sublayer" near the wall disappears. Thus, more exact results near the wall are obtained, and the equation for ν_T is of the form [23]:

$$\frac{l^4 \nu_T (\tau/\rho)^2}{(\nu + \nu_T)^2} - \nu_T\left(\nu_T + \frac{\nu}{\alpha}\right)^2 + \frac{y^4}{2\alpha\rho} \frac{\partial}{\partial y}\left\{\rho\nu \frac{\partial}{\partial y}\left[\frac{\nu_T}{y^2}\left(\nu_T + \frac{\nu}{\alpha}\right)\right]\right\} = 0 \tag{1.35}$$

When Eqs. (1.35) and (1.30) do not possess the last (diffusional) term, they coincide. Once the calculation is made by (1.35), the empirical constant α differs somewhat from α used in (1.30) and amounts to $\alpha = 0.152$ [23].

These semiempirical theories may be employed alongside of purely empirical methods and allow the field of application of the empirical approach to be specified.

1.3 THE METHODS OF CALCULATING A BOUNDARY LAYER

In the intertube space of helical tube heat exchangers, the wall temperature distribution over the apparatus length and radius may be determined using different calculation methods of a wall (boundary) layer if the external flow is specified. The external flow, i.e., the flow in the flow core, is found by solving the system of equations presented in Chapters 3 and 4. In this case, a computer can be used to solve the system of equations for the boaundary layer which was analyzed in the previous section. The wall conditions outside the boundary layer (length distributions of velocity, temperature, and pressure), as well as the velocity and temperature profiles in the initial cross section of the wall boundary layer must be specified as boundary conditions. The boundary layer calculations yield the velocity and temperature fields as well as the length distributions of the wall heat flux and wall friction.

A simpler and more convenient method, that can be easily refined by experimental data (results for coefficients of heat transfer and hydraulic resistance, velocity and temperature fields), is based on the replacement of the initial system of two-dimensional equations by a system of approximate one-dimensional ones. Such a method was proposed in [23]. In this case, the one-dimensional equations can be solved using quickly converging successive approximations (two approximations are frequently enough to calculate friction and heat transfer). In converting to a one-dimensional problem, it is assumed that the form of the dimensionless velocity profiles and of other quantities in the boundary layer change rather slowly along the body in the flow.

The distribution of τ and q_0 is found by Eqs. (1.10)-(1.12), and the quantity q by expression (1.15):

$$\tau=\tau_W-\bar{\rho}\bar{u}\frac{\partial\bar{u}}{\partial x}y+\int\limits_0^y\frac{\partial\rho u^2}{\partial x}dy-u\int\limits_0^y\frac{\partial\rho u}{\partial x}dy \tag{1.36}$$

$$q_0=q_W+\int\limits_0^y\frac{\partial\rho u I_0}{\partial x}dy-I_0\int\limits_0^y\frac{\partial\rho u}{\partial x}dy \tag{1.37}$$

considering that $dp/dx = \bar{\rho}\bar{u}d\bar{u}/dx$. Neglect of the integrals on the right-hand side of these equations gives:

$$\tau^*=\tau_W-\bar{\rho}\bar{u}\frac{d\bar{u}}{dx}y \tag{1.38}$$

$$q_0^*=q_W \tag{1.39}$$

On the wall ($y=0$)

$$\tau_w = \tau^*_w;\ \left(\frac{\partial\tau}{\partial y}\right)_w = \left(\frac{\partial\tau^*}{\partial y}\right)_w;\ \left(\frac{\partial^2\tau}{\partial y^2}\right)_w = \left(\frac{\partial^2\tau^*}{\partial y^2}\right)_w \tag{1.40}$$

$$q_w = (q_0)_w = (q^*_0)_w;\ \left(\frac{\partial q_0}{\partial y}\right)_w = \left(\frac{\partial q^*_0}{\partial y}\right)_w;\ \left(\frac{\partial^2 q_0}{\partial y^2}\right)_w = \left(\frac{\partial^2 q^*_0}{\partial y^2}\right)_w \tag{1.41}$$

Hence, on the wall the quantities τ, τ^*, q_0, q^*_0, and their first and second derivatives coincide. Therefore, approximate determination of the integrals in (1.36) and (1.37) gives rise to errors only in the region far from the wall, where this error is not very substantial. This is because far from the wall the quantities u and I_0 vary slightly at large values of ν_T and a_T in this region. It may be approximately assumed that u/u and $(I_0/I_w)/(\bar{I}_0 - I_w)$ depend only on y/ϑ and y/Θ, where ϑ and Θ are the momentum and energy loss thicknesses, respectively. Then, the derivatives of x with respect to different quantities under the integral signs in (1.36) and (1.37) may be expressed in terms of y-derivatives, for example,

$$\frac{\partial u}{\partial x} \approx u\frac{1}{\bar{u}}\frac{d\bar{u}}{dx} - y\frac{du}{dy}\frac{1}{\vartheta}\frac{d\vartheta}{dx},\ \text{etc}\ \frac{1}{\vartheta}\frac{d\vartheta}{dx}\ \text{and}\ \frac{1}{\Theta}\frac{d\Theta}{dx}$$

may be found from the integral momentum and energy relationships for a boundary layer. Removing from the integration signs some mean values of density and cancelling them, when the velocity profiles and I_0 are assumed to be similar, i.e., $u/\bar{u} \approx (I_0 - I_w)/(\bar{I}_0 - I_w) \approx (y/\delta)^n$, where δ is the boundary layer thickness, we may find approximate values of the integrals in Eqs. (1.36) and (1.37) and have:

$$\tau \approx \tau_w\left[-\left(\frac{u}{\bar{u}}\right)^{2+\frac{1}{n}}\right] - y\bar{\rho}\bar{u}\frac{d\bar{u}}{dx}\left[1-\left(\frac{u}{\bar{u}}\right)^2\right] \tag{1.42}$$

$$q_0 \approx q_w\left[1-\left(\frac{I_0 - I_w}{\bar{I}_0 - I_w}\right)^{2+\frac{1}{n}}\right] \tag{1.43}$$

When Eqs. (1.16) and (1.17) are supplemented both with δ from (1.42) and $q = q_0 - u\tau$, where q_0 is determined by (1.43) and with ν_T and a_T, they form a system of two ordinary differential equations of the first order with two unknown functions of y, i.e., u and I, that can be easily solved numerically with the boundary conditions at $y = 0$, $u = 0$, and $I = I_w$. Thus, when I_w,

u, I_0, du/dx, q_w, and τ_w are known, the profiles of u, I, and other quantities can be calculated by integrating expressions (1.16) and (1.17) with the aid of (1.42), (1.43), and the above formulas for ν_T and a_T. The value of n in (1.42) and (1.43) is prescribed from experimental data.

When this approximate calculation method for a boundary layer is used, the following dimensionless quantities are introduced:

$$a = \frac{\tau_w}{\rho_x \bar{u}^2} \tag{1.44}$$

$$a_m = \frac{q_w}{\rho_x \bar{u} j_0} \tag{1.45}$$

$$\left.\begin{aligned} &z = \frac{\bar{u}\rho_x\vartheta}{\mu_x a};\ z_m = \frac{\bar{u}\rho_x\Theta}{\mu_x a_m},\ d\,\mathrm{Re}_x = \frac{\rho_x\bar{u}\,dx}{\mu_x} \\ &H = \frac{1}{(\rho\vartheta)}\int_0^\delta \rho\left(1 - \frac{u}{\bar{u}}\right)dy;\ H_\rho = \frac{1}{(\rho\vartheta)}\int_0^\delta \rho\left(1 - \frac{\bar{\rho}}{\rho}\right)dx \\ &H_m = \frac{1}{(\rho\Theta)}\int_0^\delta \rho\,\frac{\bar{I}_0 - I}{j_0}\,dy. \end{aligned}\right\} \tag{1.46}$$

where

$$(\rho\vartheta) = \int_0^\delta \rho\frac{u}{\bar{u}}\left(1 - \frac{u}{\bar{u}}\right)dy,\ (\rho\Theta) = \int_0^\delta \rho\frac{u}{\bar{u}}\,\frac{\bar{I}_0 - I}{\bar{I}_0 - I_w}\,dy \tag{1.47}$$

$\rho_x = \rho_x(x)$ is some characteristic ("mean") density in the considered cross section of a boundary layer; μ_x = const is some mean (over the boundary layer cross section and along the entire body in the flow) value of the dynamics viscosity of the medium; and j_0 is the effective enthalpy head

$$\bar{j}_0 = \bar{I} + \psi\frac{\overline{u^2}}{2} - I_w \tag{1.48}$$

where ψ is the recovery factor. The quantities H, H_Q and H_m depend both on the form of the velocity and enthalpy profiles and on the form of the state equations for the medium.

The next stage is to select mean values of a medium density, ρ_x, and viscosity, μ_x, and a mean Prandtl number so that the relationships between the dimensionless parameters become approximately the same as those for a fluid with constant properties. For a constant property fluid, only $\bar{u}$, $d\bar{u}/dx$, $I-I_w$, ϑ, Θ, and the medium properties ρ, ν, and a must be specified to calculate the velocity profiles, $I\text{-}I_W$, as well as τ_W, q_W, and $I_{stagn}\text{-}I_W$ using the above methods. Then, the relationships between the dimensionless parameters for this flow will be of the form:

$$u \quad u\left(z, \frac{\vartheta}{\bar{u}} \frac{d\bar{u}}{dx}\right)$$

$$H \quad H\left(z, \frac{\vartheta}{\bar{u}} \frac{d\bar{u}}{dx}\right)$$

$$\psi = \psi\left(z, z_m, \frac{\vartheta}{\bar{u}} \frac{d\bar{u}}{dx}, \Pr\right) \tag{1.49}$$

$$u_m = u_m\left(z, z_m, \frac{\vartheta}{\bar{u}} \frac{d\bar{u}}{dx}, \Pr\right)$$

$$H_m = H_m\left(z, z_m \frac{\vartheta}{\bar{u}} \frac{d\bar{u}}{dx}, \Pr\right)$$

These formulas may be found both numerically and empirically.

The experimental relations for the calculation of heat transfer and hydraulic resistance in a helical tube bundle in the form of (1.49) will be derived below. The calculation results for a variable-property medium may be represented using the same formulas in selecting ρ_X, μ_X, and $\Pr_X$. In [23] it is shown that when the medium properties change monotonically, the formula for the case of a constant-property medium is also valid for calculating a variable-property medium, assuming that

$$\rho_x = \rho_m^{0,82}\rho_*^{0,18}; \ \mu = \mu_m; \ \Pr_x = \Pr_m \tag{1.50}$$

under the following characteristic parameters:

$$I_m = \frac{\bar{I}_0 + I_W}{2} - \left(\frac{\bar{u}}{2}\right)^2 \Big/ 2$$

$$I_* = \frac{3\bar{I}_0 + I_W}{4} - \left(\frac{3\bar{u}}{4}\right)^2 \Big/ 2 \tag{1.51}$$

A length variation of arguments z and z_m in the equations of form (1.49) is found from the integral momentum and energy relations in [23]:

$$\frac{dz}{d\,\mathrm{Re}_x} + \frac{z}{u}\frac{du}{d\,\mathrm{Re}_x} + (1 + H - H_\rho)\frac{z}{\bar{u}}\frac{d\bar{u}}{d\,\mathrm{Re}_x} + \frac{z}{R}\frac{dR}{d\,\mathrm{Re}_x} = 1 \tag{1.52}$$

$$\frac{dz_m}{d\,\mathrm{Re}_x} + \frac{z_m}{u_m}\frac{du_m}{d\,\mathrm{Re}_x} + \frac{z_m}{\bar{j}_0}\frac{d\bar{j}_0}{d\,\mathrm{Re}_x} + \frac{z_m}{R}\frac{dR}{d\,\mathrm{Re}_x} = 1 \tag{1.53}$$

For the one-dimensional case, R = const and $d/\mathrm{Re}_x = \rho_x \bar{u} dx/\mu_x$. Relations (1.52) and (1.53) fully close the system of equations used to calculate a boundary layer according to this method. The order of a calculation procedure using this approximate method is presented in [23]. Some recommendations on this method, when used to calculate heat transfer in helical tube bundles, will be detailed in the next sections of this book.

1.4 THE SIMILARITY OF HEAT TRANSFER AND FLOW PROCESSES IN LONGITUDINAL FLOW PAST A HELICAL TUBE BUNDLE

Usually, when dimenionless relations for the processes of heat transfer and flow in one or another device are determined, the requirements for geometrical, kinematic, dynamic, and thermal similarity must be satisfied. These functional relationships between the quantities describing the investigated phenomena are found, using similarity and dimensional theories, in the form of a relationship between dimensionless quantities, or similarity numbers [45]. In a number of cases, it is possible, by combining similarity theory with physical and experimental considerations, to yield a minimum of dimensionless parameters that satisfy the main effects in the most convenient form. This facilitates the design and performance of experiments.

First, let us consider the problem of determining the form of the dimensionless equation for the hydraulic resistance coefficient in a helical tube bundle. For this purpose, an analysis should be made of the forces that affect the fluid flow and characterize its motion. This problem may be represented as follows. A helical tube bundle and a part of its cross section, through which a fluid is passing, are determined either by assigning a cross-sectional area F_f or a characteristic linear size. An equivalent diameter, found in terms of a wetted wall perimeter Π_{wet}, $d_{eq} = 4F_f/\Pi_{wet}$, may be taken as a characteristic linear size of complex-geometry channels. From experience and experiment it is known for noncircular channels and rod bundles that the inclusion of d_{eq} usually results in a better correlation of experimental data. The specific features at the inlet and outlet are not taken into account. The fluid motion is steady. The effect of flow nonisothermity and gas compressibility will not be considered. The property of inertia is characterized by a density ρ, and that of viscosity by a viscosity coefficient μ. The effect of a test body temperature on ρ and μ is taken into account by introducing known relations. The fluid motion is determined either by assigning a fluid flowrate or a mean mass (mean with respect to a cross section) velocity u_{mean}. The action of an inertia field of centrifugal forces in a bundle created with a fluid flowing in spiral channels is taken into consideration in geometrically similar bundles when the conditions s/d_{eq} = idem, d/d_{eq} = idem, s/d = idem, etc. are satisfied. Then, the system of parameters d_{eq}, ρ, μ, u_{mean} is obtained at three independent main units of measurement: length, in m; mass, in kg; time, in s. These can be the basis of one dimensionaless combination:

$$\mathrm{Re}_d = \frac{u_{mean} d_{eq} \rho}{\mu} \tag{1.54}$$

The dimensionless resistance coefficient $\xi = |dp_0/dx| 2d_{eq}/\rho u^2_{mean}$ for a helical tube bundle is a function of Re_d $\xi = \xi(\mathrm{Re}_d)$.

If geometrically nonsimilar bundles of helical tubes with different porosities m, number of tubes n, with a considerable difference in s/d, s/d_{eq}, h/d_{eq}, d/d_{eq}, etc., where h is the minimum size of the oval, are examined, then the criterial relation will be of the form:

$$\xi = \xi\left(\mathrm{Re}_d, \frac{d}{d_{eq}}, \frac{s}{d_{eq}}, \frac{h}{d_{eq}}, n \ldots\right)$$

It is very difficult to obtain such relations experimentally.

The approximate hydrodynamic similarity, when a fluid is flowing in a helical tube bundle with no geometrical similarity, may probably be ensured if the maximum acceleration values of a centrifugal force field, $g_{c,max}$, are used as a parameter that characterizes the specific features of the bundle flow.

Therefore, for stabilized steady turbulent incompressible liquid flow under nearly isothermal conditions, the following system of parameters is obtained: d_{eq}, ρ, μ, u_{mean}, $g_{c,max}$. From this system it is possible to derive two dimensionless combinations, Reynolds number (1.54) and $\mathrm{Fr}_c = u^2_{mean}/g_{c,max} d_{eq}$, responsible for the impact of centrifugal forces on the bundle flow. The coefficient ξ depending on the above parameters is a function of the characteristic Re_d and Fr_c: $\xi = \xi(\mathrm{Re}_d, \mathrm{Fr}_c)$.

If the fluid flow in spiral tube channels is assumed to swirl according to the rigid body law $u_\tau r^{-1} = \mathrm{const}$, then a maximum value of a tangential velocity component $u_{\tau max}$ will be related to u_{mean} by $u_{\tau max} = \pi d u_{mean}/s$. Since the acceleration in a centrifugal force field is equal to $g_{c,max} = 2u^2_{\tau max}/d$, the criterion for a centrifugal force field in a spirally twisted tube bundle may be reduced to the form $Fr_c = s^2/2\pi^2 d\, d_{eq}$, where it is only a function of the geometrical sizes of the helical tubes and the bundle. Thus, the Fr_c number or the modified number

$$Fr_m = \frac{s^2}{dd_{eq}} \tag{1.55}$$

is a complex geometrical characteristic of a helical tube bundle that specifies the bundle flow. Then, the criterial equation for ξ in geometrically nonsimilar helical tube bundles will be of the form:

$$\xi = \xi(Re_d,\ Fr_m) \tag{1.56}$$

This relationship must be determined from experiment.

In addition to the above five parameters, the coordinates of spatial points must be introduced to study the velocity field distributions. Then, the stabilized flow length will be governed by the following criterial relation:

$$\frac{u}{\bar{u}} = f\left(Re_d,\ Fr_m,\ \frac{y}{d_{eq}}\right) \tag{1.57}$$

where $\bar{u}$ is the maximum flow velocity.

Since the secondary flow circulation developing in a centrifugal force field influences the flow core and the thin boundary layer on the helical tubes, whose effective thickness of δ must depend on Fr_M, it may be assumed that when teh quantity δ is included as a characteristic bundle size, we may obtain the universal criterial relations:

$$\xi = \xi(Re_\delta) \tag{1.58}$$

$$\frac{u}{\bar{u}} = f\left(Re_\delta,\ \frac{y}{\delta}\right) \tag{1.59}$$

where

$$Re_\delta = \frac{u_{mean}\rho\delta}{\mu} \tag{1.60}$$

As shown below, inclusion of the quantity δ enables one to present experimental data on ξ, Nu, velocity, and temperature profiles in a form where the validity of the hydrodynamic heat transfer theory is clearly seen, and it proves possible to apply boundary layer calculation methods to the present flow case.

From Eqs. (1.58) and (1.59) it follows that the influence of Fr_M on ξ and $u/\bar{u}$ manifests itself in terms of the integral geometric characteristic of a bundle, namely, a boundary layer thickness δ entering Re_δ. This simplifies the relations for the hydraulic resistance coefficient and the velocity profile. However, in this case, it is also necessary to determine from experiment the relationship:

$$\frac{\delta}{d_{eq}} = \varphi(Fr_m) \tag{1.61}$$

From Eqs. (1.58), (1.59), and (1.61) it follows that, in a number of cases, it is possible to extend the modelling potentialities to yield a more clear understanding of a physical pattern of heat transfer and flow processes in a helical tube bundle and to work out more general methods of calculating friction and heat transfer.

For geometrically nonsimilar helical tube bundles, when the quantity g_c is included the problem of finding the form of the criterial equation for heat transfer is also simplified.

The unknown thermal conductivity

$$\alpha = \frac{q}{T_w - T_f} \tag{1.62}$$

is a function of the parameters that characterize the fluid motion - d_{eq}, ρ, μ, u_{mean}, and g_c - and of the parameters for fluid thermal conductivity λ and heat capacity c_p. Then, we have the following system of parameters - α, d_{eq}, ρ, μ, u_{mean}, g_c, λ, c_p - at four independent basic units of measurement: meter, kilogram, second, degree. Based on this system, it is possible to make four dimensionless combinations:

Nusselt number $\quad Nu = \dfrac{\alpha\, d_{eq}}{\lambda},$

Prandtl number $\quad Pr = \dfrac{\mu\, c_p}{\lambda},$

Reynolds number, Re (1.54), and

Froude number, Fr_M (1.55)

The unknown dimensionless relation for the stabilized flow length under nearly isothermal conditions will be of the form:

$$Nu_d = Nu(Re_d,\ Fr_M,\ Pr) \tag{1.63}$$

Inclusion of the quantity δ, determined by expression (1.61) as a characteristic size, enables one in this case to obtain the following functional relationship

$$Nu_\delta = Nu(Re_\delta,\ Pr) \tag{1.64}$$

where

$$Nu_\delta = \frac{\alpha\delta}{\lambda} \tag{1.65}$$

To take into account the specific features of the flow at the bundle inlet, criterial relations (1.56), (1.50), (1.63), and (1.64) should be supplemented with a dimensionless longitudinal coordinate at the bundle inlet x/d_{eq} or x/δ as a characteristic parameter. The flow nonisothermity effect on heat transfer and hydraulic resistance is usually allowed for by including a temperature factor T_w/T_f as a characteristic group into the criterial equation.

In addition to the parameters d_{eq}, ρ, μ, u_{mean}, λ, and c_p, the transverse y-coordinate normal to the tube wall must be included to obtain a criterial relation that describes the temperature fields in the boundary layer at the stabilized flow length. Then, we have the relationship

$$\frac{T_w - T}{T_w - \bar{T}} = \psi\left(Re_d,\ Fr_M,\ Pr,\ \frac{y}{d_{eq}}\right) \tag{1.66}$$

$$\frac{T_w - T}{T_w - \bar{T}} = \psi\left(Re_i,\ Pr,\ \frac{y}{d_\delta}\right) \tag{1.67}$$

where $\bar{T}$ is the fluid temperature at the external edge of the boundary layer.

The form of the criterial relations that was established using similarity and dimensional theories and which describes heat transfer and friction processes differs from the one for circular tubes. This difference is determined by the effect of the centrifugal force field on the flow in a helical tube bundle, which creates an extra dynamic similarity number (1.55) that either directly enters the functional relations or enters indirectly through the characteristic size δ determined by expression (1.61). The representation of experimental data on heat transfer and hydrodynamics in helical tube bundles in the form of the proposed criterial relations extends the modelling potentialities of heat transfer and flow dynamics and diminishes the number of experimental studies necessary to establish these relations.

1.5 PROBLEM AND ITS SOLUTION METHODS

This book is concerned with the problem of designing and calculating compact heat exchangers intended for aviation engineering, in which heat and momentum transfer is enhanced by flow swirling in complex-geometry channels formed by helical tubes. Comprehensive studies of the turbulent flow structure, velocity, and temperature fields, heat transfer, effective turbulent thermal conductivity, and viscosity coefficients for different heat exchanger designs must be carried out in order to calculate thermal and hydraulic characteristics with regard to interchannel mixing of the heat carrier in the intertube space of the heat exchanger.

Considering that the flow in helical tube bundles is complex in nature and that it is impossible to strictly solve this problem, it is advisable to use the mathematical apparatus for analyzing different physically grounded flow models which are based on experimental studies. In this monograph much attention is paid to obtaining effective transfer coefficients as a function of characteristic similarity numbers in order to complete systems of differential equations that describe the flow within the framework of the adopted models. The methods for approximate closure of the equations for different flow regions (in the flow core and in the wall layer) are experimentally supported on the experimental model set-ups. In this case, much attention is concentrated on extending the modelling potentialities and on extending the data, obtained under one set of conditions, e.g., for an incompressible fluid, to the calculation of helical tube bundles under other conditions, e.g., for a variable property uniform medium. In obtaining the criterial relations for calculating heat transfer, hydraulic resistance, effective viscosity coefficient, thermal conductivity, and diffusion coefficient, similarity and dimensional theories for geometrically nonsimilar helical tube bundles were developed and the original investigation methods were elaborated, thus making it possible to distinguish the considered effects more clearly. Since the thermal and hydraulic characteristics in helical tube bundles were studied for the first time, it was very important to substantiate the relationship between these characteristics and the characteristic similarity numbers. This was done by thermoanemometric studies of averaged and pulsational motions in a bundle as well as by analysis of energy turbulence spectra. Consequently, the reliability of the obtained results increased.

In this book, the experimental and theoretical methods used to solve this problem supplement each other. Thus, the theoretical methods for investigating and calculating heat carrier temperature fields and flow velocities in helical tube bundles are supported by correspondence between the predicted and experimental results on the model set-ups. The experimental data enabled developing the method of calculating the efficiency of helical tube heat exchangers having different designs. When boundary layer methods were used, it became possible to obtain experimental data on heat transfer and hydraulic resistance in a universal form.

The experimental methods adopted in this book permitted establishing the boundaries of swirled flow regimes in helical tube bundles, as well as the field of application of the dimensionless relations for the calculation of heat transfer, hydraulic resistance, and turbulent transfer coefficients.

CHAPTER
TWO

METHODS OF EXPERIMENTAL STUDY OF FLOWS IN HELICAL TUBE BUNDLES

2.1 THE SPECIFIC FEATURES OF EXPERIMENTAL STUDY OF FLOWS PAST SPIRAL SURFACES

Fluid flow in the intertube space of a helical tube heat exchanger is very complex in nature. Thus, known methods of experimental investigation of heat transfer, transport properties, and flow structure that are used to study these processes in round tubes or in flat channels, or in a boundary layer and in some other flow cases cannot therefore be directly applied to the case of flow in helical tube bundles.

For example, due to the constrained volume of the intertube space, the higher turbulence level, and the broader frequency spectrum of energy-containing vortices as compared to a straight circular or flat channel, use of a hot-wire method for studying the flow structure in a helical tube bundle has required the design of a special hot inclined-wire anemometer with a Pitot tube, refinement of the equation for a hot-wire anemometer, and adjustment of the Doppler DISA equipment for a substantially broader frequency spectrum (50000 Hz), as compared to the frequency tuning in a circular channel (20000 Hz). At the same time, since there were no data on the flow structure in helical tube bundles it was necessary to thoroughly calibrate anemometers and other equipment as well as to check the reliability of the data obtained from experiments in circular tubes and to compare them with the well-known results of different authors.

This characteristic of the experimental study of spirally swirled flows in helical tube bundles has also been applied in determining the characteristics of heat transfer and other transport processes. When a total pressure gauge was used, much attention was paid to its geometrical size. In this case, the gauge should have a low sensitivity to flow angularity. This was achieved by choosing a hole value equal to d_{in}/d_{out} = 0.834 and a shape of the total pressure gauge nose which was not rounded off near a hole. In this case, the values of total pressures remained invariable at angles equal to $\varphi \approx$ 0.03834 π, which are larger than those needed for the investigation of helical tube bundles. Thus, when the Pitot tube was located on a coordinate screw mechanism in a manner such that the tube axis was normal to the bundle cross section, a total pressure, p_{tot}, was practically measured at each flow point. Static pressure taps in the cross section where p_{tot} was measured were equidistant on the periphery of the shell of the experimental section. The dynamic head $\rho V^2/2 = p_{tot} - p$ measured by a U gauge allowed a flow velocity vector V to be determined at each flow point in terms of a known density profile ρ. This velocity may be considered as a time-mean since the pulsation frequency of the measured quantities was much higher than the natural frequencies of the set-up.

The specific flow features had to be taken into account to study the hydraulic resistance coefficient for a helical tube bundle, although generally accepted experimental methods might have been used. Adjustment of the experimental set-up and procedures also had to be made in the circular tubes to improve the reliability of the experimental data. Experiments on total pressure gauges in air at Re = $2{\cdot}10^3$–$8.2{\cdot}10^4$ and $M \le 0.6$ have shown that pressure gauges are hydraulically smooth and the friction coefficient obeys Blasius' law. The set-up adjusted in this fashion and the measuring system enabled obtaining the hydraulic resistance coefficients in a helical tube bundle within a maximum error of $\pm$8%. This experimental study of hydraulic resistance coefficients was characterized by the fact that the effect of a transverse velocity component field on the readings of static pressure gauges was considered. Therefore, in these experiments the static pressure gauges were mainly mounted at places where the total velocity vector in the flow core was parallel to the generatrix of the tube shell or helical tube.

The complex flow pattern associated with the contacts of adjacent tubes, and a periodic interaction of helical flows in spiral channels along the bundle which contact one another or through channel flow, may also promote a periodic change in the heat transfer coefficient along the helical tube. This characteristic property must be allowed for in experimental studies dealing both with the location of thermocouples measuring the helical tube wall temperature and with the development of a special moving thermocouple probe measuring the wall temperature distributions along the tube at different tube perimeter points. This probe was located inside the helical tubes and the thermocouple junctions were pressed into the tube wall with a special instrument.

The distinctive feature of the experimental heat transfer study was the method of measuring the flow temperature at the outlet of a helical tube bundle having an axisymmetric air efflux. In this case, measurements were made not only of the mean mass (bulk) temperature but of the temperature field, in terms of which a mean mass temperature at the bundle outlet was calculated. In a number of cases, the local heat transfer coefficient with a nonuniform heat release was determined through the flow temperature measured at the external boundary of the wall layer.

The specific features of the experimental study of transport flow properties were stipulated by a choice of certain flow models. Thus, when a homogenized flow model was used, the effective turbulent diffusion coefficients for a helical tube bundle were found as if we considered that a homogenized medium with distributed sources and hydraulic resistance was flowing in a cylindrical channel equal to the tube shell diameter. In this case, using the known method of diffusion from a fixed point heat source in a constant velocity field, it was necessary to make corrections for the real behavior of helical tubes. This required a diffusion source to move and velocity and temperature fields to be measured in one outlet cross section. Thus, a diffusion source size was chosen, assuming that the properties of the considered flow and indicator gas were identical. If the method of heat diffusion from a linear source was used directly, then the specific features of developing temperature nonuniformity would not have been taken into account because the main transfer mechanisms would take effect only when a temperature wake developed up to a radial size including 7-19 tubes. This would require very large lengths of an experimental section. As a result, this method was modified and a system of linear sources was used as a heat diffusion source. In what follows, the number of heated helical tubes varied from 7 to 37 in modelling axisymmetric and asymmetric nonuniformity of heat release.

The specific features of the experimental study of interchannel flow mixing in a helical tube bundle were also associated with an account of the impact of the swirled flow on the measured temperature and velocity fields. Measurements were made using the accepted flow model. Therefore, the temperature and velocity measurements were performed mainly outside the wall layer. Here, the effect of the heating zone shape was taken into consideration, and the tubes for this zone were chosen with regard to their ohmic resistance. Electric insulation is required between the heating zone and the nonheated tubes of a bundle to implement the method of diffusion from a system of linear heat sources.

Specific features of the experimental study of heat transfer and flow dynamics will be detailed in the sections dealing with these processes.

2.2 METHODS OF STUDYING VELOCITY FIELDS AND THE TURBULENT FLOW STRUCTURE

Velocity fields and the turbulent flow structure in helical tube bundles should be studied on large-scale models where the oval size of the tubes is a maximum. In this way, a total velocity vector may be measured by a total pressure gauge not only in the flow core but also in the wall layer of the helical tubes [8, 9, 15, 16], and velocity vector components may be measured by a constant temperature anemometer. Usually, use was made of total pressure gauges having a small diameter and a sensitivity to the flow angles of up to $\varphi \approx 0.0834\pi$. A U tube differential manometer was used to measure dynamic heads

$$\frac{\rho V^2}{2} = p_{tot} - p \tag{2.1}$$

The total pressure gauge and other probes were mounted on a traversing coordinate mechanism.

The methods used to study total velocity vector components in the three-dimensional flow in a helical tube bundle were based on the use of a small rotating hot inclined-wire anemometer of original design. This anemometer allowed measurement in the constrained intertube space of a heat exchanger (Fig. 2.1) [9]. When a single-wire anemometer

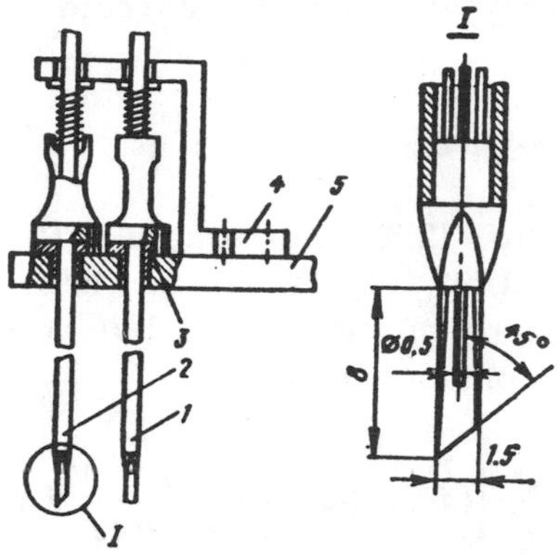

Figure 2.1 Hot inclined and straight-wire anemometers mounted on one holder: *1, 2)* body; *3)* centering bushing; *4)* camp; *5)* holder

was used, the difficulties associated with the difference in characteristics of different wires in a multi-wire anemometer were also obviated, and the measuring accuracy was improved. Measurements in two mutually perpendicular planes were made by swinging the anemometer vertically. The wire inclination relative to the anemometer axis was 45°. The anemometer and the total pressure gauge used to measure velocity vectors were mounted on a coordinate mechanism.

Measurements and calculations of averaged and pulsating velocity components of the three-dimensional flow were made separately, assuming that there were no restrictions on the angle characterizing the velocity vector direction and that there existed a high turbulence flow level. The governing equation for the hot-wire anemometer that relates an instantaneous velocity vector *V*:

$$V^2 = u^2 + v^2 + w^2 \tag{2.2}$$

and an instantaneous wire voltage drop *E* had the form:

$$E^2 = E_0^2 + B(\rho V)^c (\sin^2\varphi + k^2 \cos^2\varphi)^{c/2} \tag{2.3}$$

where *B, c,* and k^2 are a weak function of velocity. Equation (2.3) for an inclined wire is obtained by the empirical relation [9]:

$$\mathrm{Nu}\left(\frac{T_w + T_f}{2T_f}\right)^{-0.17} = a + B\,\mathrm{Re}^c \tag{2.4}$$

where the velocity entering the Reynolds number is determined by the formula:

$$V_{ef}^2 = V_{\varphi-90^\circ}^2 (\sin^2\varphi + k^2\cos^2\varphi) \tag{2.5}$$

Differentiating Eq. (2.3) and squaring the expression obtained upon differentiation of (2.3), we arrive at the governing equation for the calculation of turbulent stresses and averaged velocity vector components [9]

$$\left(\frac{2E}{E^2 - E_0^2}\right)^2 dE^2 = \frac{c^2}{V^2}\left[(dV)^2 + 2\frac{1-k^2}{\text{tg}\,\varphi + k^2\,\text{ctg}\,\varphi} V\,dV\,d\varphi + \left(\frac{1-k^2}{\text{tg}\,\varphi + k^2\,\text{ctg}\,\varphi}\right)^2 (V\,d\varphi)^2\right] \tag{2.6}$$

Since

$$E = \bar{E} + e',\quad u = \bar{u} + u'$$
$$v = \bar{v} + v',\quad w = \bar{w} + w' \tag{2.7}$$
$$\varphi = \bar{\varphi} + \varphi'$$

expression (2.6) may be written as:

$$\left[\frac{\bar{E}^2 - E_0^2 + 2\bar{E}e' + \overline{e'^2}}{B\rho^c}\right]^{2/c} = (\bar{u}^2 + \bar{v}^2 + \bar{w}^2 + \overline{u'^2} + \overline{v'^2} + \overline{w'^2} + 2\bar{u}u' + 2\bar{v}v' + 2\bar{w}w')\left[\sin^2(\bar{\varphi} + \varphi') + k^2\cos^2(\bar{\varphi} + \varphi')\right] \tag{2.8}$$

Equation (2.8) enables establishing the relationship between the experimentally measured quantities $\bar{E}$ and $\sqrt{\overline{e'^2}}$ and the initial values of the velocity vector components. Since the quantity $\sqrt{\overline{e'^2}}/(E-E_0) < 0.1$ (from experiment), the LHS of Eq. (2.8) may be expanded into a series. Then, after the terms of higher orders are averaged and discarded, we obtain

$$\left[\frac{\bar{E}^2 - E_0^2 + 2\bar{E}e' + \overline{e'^2}}{B\rho^c}\right]^{2/c} = \left(\frac{\bar{E}^2 - E_0^2}{B\rho^c}\right)^2\left(1 + \frac{2\overline{e'^2}}{c(\bar{E}^2 - E_0^2)}\right) \times\left[1 + \left(\frac{2}{c} - 1\right)\frac{2\bar{E}^2}{\bar{E}^2 - E_0^2}\right] \tag{2.9}$$

The RHS of Eq. (2.8), with regard to infinitesimal fluctuations φ' and neglecting terms higher than the 3rd order, may be reduced to the form:

$$(\sin^2\bar{\varphi}+k^2\cos^2\bar{\varphi})\,[(\overline{u^2}+\overline{v^2}+\overline{w^2})\,(1+k^2\overline{\varphi'^2}\sin^2\bar{\varphi}+\overline{\varphi'^2}\cos^2\bar{\varphi})$$
$$+\bar{u}^2+\bar{v}^2+\bar{w}^2+2(1-k^2)^2\frac{\sin\bar{\varphi}\cos\bar{\varphi}}{\sin^2\bar{\varphi}+k^2\cos^2\bar{\varphi}}$$
$$\times(\bar{u}\overline{u'\varphi'}+\bar{v}\overline{v'\varphi'}+\bar{w}\overline{w'\varphi'})] \quad (2.10)$$

If we designate the multiplier in brackets, expression (2.10), as Q, as experiments have shown, this multiplier practically does not depend on the angle $\bar{\varphi}$. Then, Eq. (2.8) will be of the form:

$$F_i=(\sin^2\bar{\varphi}_i+k^2\cos^2\bar{\varphi}_i)\,QB^{2/c}\rho^2 \quad (2.11)$$

where

$$F_i=(\bar{E}_i^2-E_0^2)^{2/c}\left\{1+\frac{2\overline{e'^2}}{c\,(\bar{E}_i^2-E_0^2)}\left[1+\left(\frac{2}{c}-1\right)\frac{2\bar{E}_i^2}{\bar{E}_i^2-E_0^2}\right]\right\} \quad (2.12)$$

The subscript "i" stands for the position of the anemometer relative to the planes where measurements are being made. Numbers *1* and *3* represent measurements in the xy-plane, while *2* and *4*, are in the xz-plane (Fig. 2.2). The LHS of Eq. (2.11) covers the quantities either directly measured by experiment, E_i, $\sqrt{\overline{e'^2}}$, or determined from a calibration of c and E_0. The term $Q{\cdot}B^{2/c}\,\rho^2$ in the RHS of Eq. (2.11) does not depend on $\bar{\varphi}$ and may be eliminated by making measurements in two mutually
perpendicular planes:

$$F_1/F_3=F_{xy}=\frac{\sin^2\bar{\varphi}_1+k^2\cos^2\bar{\varphi}_1}{\sin^2\bar{\varphi}_3+k^2\cos^2\bar{\varphi}_3} \quad (2.13)$$

$$F_2/F_4=F_{xz}=\frac{\sin^2\bar{\varphi}_2+k^2\cos^2\bar{\varphi}_2}{\sin^2\bar{\varphi}_4+k^2\cos^2\bar{\varphi}_4} \quad (2.14)$$

The angles $\bar{\varphi}_i$ may be expressed in terms of the velocity vector components (Fig. 2.3) as:

$$\cos\bar{\varphi}_{1,3}=(\bar{u}\cos\alpha_{1,3}-\bar{v}\sin\alpha_{1,3})/\bar{V} \quad (2.15)$$

$$\cos\bar{\varphi}_{2,4}=(\bar{u}\cos\alpha_{2,4}-\bar{w}\sin\alpha_{2,4})/V \quad (2.16)$$

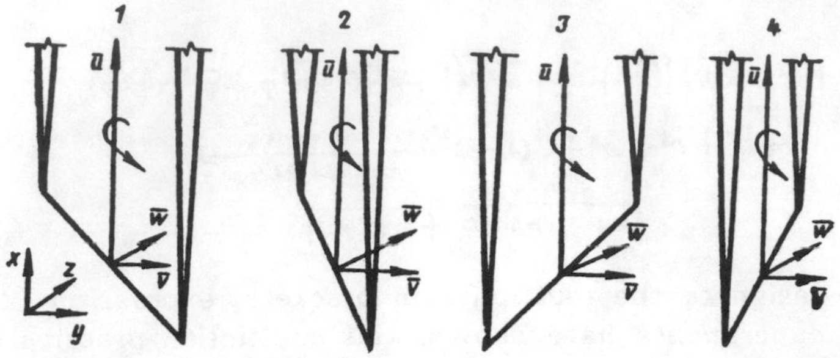

Figure 2.2 Relative location of inclined-wire anemometer and velocity vector components with rotation of the inclined-wire anemometer in two mutually perpendicular planes: *1, 3*) position in a *XOY*-plane; *2, 4*) position in a *XOZ*-plane

Then, eliminating the angles φ_i from expressions (2.13) and (2.14) and considering (2.2), we arrive at a system of the equations used to calculate the average velocity vector components. This system includes, together with (2.2), the following equations:

$$\overline{V}^2=(1-k^2)\left[\frac{F_{xy}+1}{F_{xy}-1}\overline{u}\,\overline{v}-\frac{1}{2}(\overline{u}^2-\overline{v}^2)\right] \tag{2.17}$$

$$\overline{V}^2=(1-k^2)\left[\frac{F_{xz}+1}{F_{xz}-1}\overline{u}\,\overline{v}-\frac{1}{2}(\overline{u}^2-\overline{w}^2)\right] \tag{2.18}$$

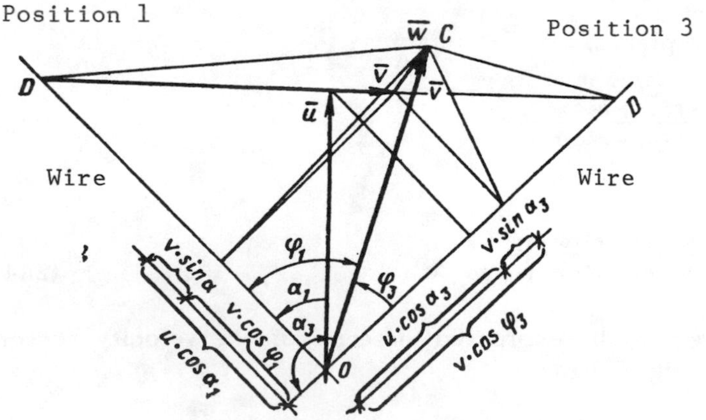

Figure 2.3 Velocity vector components versus the anemometer location

The velocity vector modulus $\bar{V}$ is determined, independent of the DISA measurements, by means of the total pressure gauge, which is insensitive to flow angles less than $\pm$ 0.0834 π.

The equation for calculating turbulent stresses due to pulsations may be obtained by establishing a relationship between the differentials dV, $d\varphi$ and the velocity vector components. Upon their substitution into Eq. (2.6) and appropriate manipulations, we have:

$$\left(\frac{2\bar{E}}{\bar{E}^2-E_0^2}\right)\overline{e'^2}=\frac{c^2}{\bar{V}^2}\left(k_{11}\overline{u'^2}+k_{12}\overline{u'v'}+k_{13}\overline{u'w'}+k_{22}\overline{v'^2}\right.$$
$$\left.+k_{23}\overline{v'w'}+k_{33}\overline{w'^2}\right) \qquad (2.19)$$

The coefficients k_{11}, k_{12}, k_{13}, k_{22}, k_{23}, and k_{33} in expression (2.19) for different positions of the anemometer in the xy- and xz-planes were calculated by a formula from [60].

Finally, we have a system of four equations in the form of (2.19) which incorporate 6 unknowns: $\overline{u'^2}$, $\overline{v'^2}$, $\overline{w'^2}$, $\overline{u'v'}$, $\overline{u'w'}$, $\overline{v'w'}$. Two unknowns must be eliminated or determined in another way to solve the above system. This may be done as follows. From preliminary experiments it was determined that in all the cases considered the coefficient k_{33} was, by an order of magnitude, less than the rest. Hence, the term $k_{33}\overline{w'^2}$ in Eq. (2.19) weakly influences the accuracy of the latter [9]. Thus, it is obvious that to a first approximation, $\overline{w'^2}$, entering the system of the equations in the form of (2.9), may be calculated from the relation $w'^2/u'^2 = f(y/\delta)$ obtained using predicted data [3]. A longitudinal pulsational velocity component may be determined from the straight-wire anemometer measurements. In this case, calculation can be made by the relation:

$$\overline{u'^2}=\frac{\bar{u}^2}{c^2}\left(\frac{2\bar{E}}{\bar{E}^2-\bar{E}_0^2}\right)^2\overline{e'^2} \qquad (2.20)$$

Thus, four equations in the form of (2.9) together with equation (2.20) are enough to determine five unknowns: $\overline{u'^2}$, $\overline{v'^2}$, $\overline{u'w'}$, $\overline{u'v'}$, $\overline{v'w'}$. As mentioned above, the quantity $\overline{w'^2}$ is found approximately using experimental data [3]. Formula (2.20) is a particular case of Eq. (2.6) at $\varphi = \pi/2$ and may be reduced to the form:

$$\frac{\sqrt{\overline{u'^2}}}{\bar{u}}=\frac{\sqrt{\overline{e'^2}}}{c}\left(\frac{2\bar{E}}{\bar{E}^2-E_0^2}\right) \qquad (2.21)$$

The constants E_0, c and k^2 are determined from calibration experiments. The quantity E_0 was found when the anemometer was coated with a hood and, consequently, $\bar{u}=0$. The quantities c and k^2 were determined over a velocity range typical of helical tube bundles, unlike some previous works which assumed k^2=const. The power at c was determined by the Colly law for a constant-superheated wire in a constant-temperature flow:

$$\overline{E}^2-E_0^2=B\,(\rho\bar{u})^c \tag{2.22}$$

Equation (2.3) was employed to find the angular sensitivity coefficient k^2.

Anemometers were calibrated in a circular tube 90 mm in diameter and more than 4 m in length. The tube outlet was provided with Vitoshinsky's nozzle promoting a uniform velocity field [9].

The measuring accuracy of a longitudinal component and total velocity vector was determined to have a rms error of no more than 3%. The rms measuring error of the tangential and radial averaged velocity components and of the longitudinal velocity component did not exceed 10%.

The above method of studying the flow structure enables one to measure the averaged velocity vector components when the transverse velocity components constitute a considerable part of the longitudinal velocity component as well as when the turbulence level of the flow is high.

2.3 METHODS OF STUDYING CROSS MIXING OF THE HEAT CARRIER

Different methods may be adopted to investigate heat carrier mixing in a helical tube bundle. Two methods were utilized in the present study. These were a heat diffusion method based on a static Lagrange description of a turbulent field to study the flow history of individual particles continuously emitted by a source, and also a diffusion method from a linear heat source based on the Euler description of a turbulent flow when the particle motion is considered as a function of time and space, relative to which, a medium moves.

The method of heat or substance diffusion from a fixed point source in the equal-velocity field has been widely adopted to measure the turbulent diffusion coefficient D_t and the turbulence intensity ε on a circular tube axis [54]. In this case, the coefficient D_t and the value of ε were determined from the limiting solutions to Taylor's equations for uniform and isotropic turbulence:

$$\frac{\partial}{\partial t}\left(\frac{1}{2}\overline{y^2}\right)=\int_{t_0}^{t}\overline{v_1(t_0)\,v_1(t')}\,dt' \tag{2.23}$$

which for small and large diffusion times are, respectively, of the form:

$$\overline{y}^2=\frac{v_1^2}{u^2}x^2 \tag{2.24}$$

$$\overline{y^2}=2\frac{D_t}{u}(x-x_0) \tag{2.25}$$

The straight line in (2.25) expresses the constancy of the diffusion coefficient and is the asymptotic equation for the experimental curve $y^2=f(x)$. Therefore, the coefficient D_t/u determined using this method is the asymptotic diffusion coefficient.

The application of this method to a helical tube bundle is warranted by the coincidence between the experimental temperature distributions of heated particle diffusion and the Gaussian distribution, since the value of a mean-static squared particle travel $\bar{y}^2$ obtained from Taylor's theory is equal to $\bar{y}^2$ calculated from the experimental distribution obeying the Gauss law.

The method of heat diffusion from a point source, when applied to a helical tube bundle, has several specific features. First, when the scale of a diffusion source is chosen, it is necessary to consider that the heated gas flowing from the source into the colder cocurrent flow must be affected by the forces typical of the flow in a helical tube bundle. Therefore, the source diameter was chosen to be equal to the maximum size of the oval of a helical tube, *d*. In this case, the properties of the flow from the source were the same as those in the helical tube bundle. This chosen scale of the diffusion source made it possible, in a small-size device, to have a flow section where the similarity of the dimensionless excess temperature profiles was observed in successively located cross sections of the nonisothermal jet in the equal-velocity field. In this case, the concept of a point diffusion source is valid.

Moreover, heated air was supplied to the central helical tube of the model to obtain a symmetric temperature profile relative to the model axis. Also, because of the structural features of the helical tube bundle, the motion reversal principle was used in the experiments. The diffusion source travelled relative to the outlet cross section where the temperature fields were measured.

A diffusion method from a system of linear heat sources was applied for the first time to helical tube bundles [12]. (Earlier, the method of heating the central tube was used to study flow mixing). The essence of this method is that a group of tubes (seven and more) is heated through ohmic resistance by passing an electric current. As a result, a nonuniformity of the heat releasing field develops and initiates temperature field nonuniformity of the heat carrier. This nonuniformity is partly smoothed due to interchannel mixing of the heat carrier. In this case, it is necessary to know how to calculate theoretically the temperature fields of the heat carrier in a bundle cross section. As a result, the coefficient D_t will be determined. A flow model is adopted to calculate temperature fields. According to this model, consideration is made of the free flow with a homogenized medium slip in the presence of the distributed sources of volumetric energy release and the hydraulic resistance within the limits allowing for a boundary layer displacement thickness δ^*. This flow is governed by the continuity equations, and the homogenization effect is taken into consideration by including the multiplier $(1-m)/m$. The coefficient D_t is determined by comparing the experimental and predicted temperature fields of the heat carrier.

The advantage of the method of diffusion from a system of linear heat sources, as against the one from a single heated tube, is that when seven tubes or more are heated, a substantially shorter bundle length is needed to develop a stabilized temperature profile. Moreover, all transport mechanisms typical of helical tube bundles take part in the developing temperature field nonuniformity of the heat carrier when heating the tube groups. As a result, the value of the desired effective diffusion coefficient may be more exactly determined from experiment.

2.4 SPECIFIC FEATURES OF AN EXPERIMENTAL SECTION DESIGN IN FLOW AND HEAT TRANSFER STUDIES

The criterial relations governing the processes of heat transfer and heat carrier mixing in helical tube bundles were established experimentally on different-size heat exchanger models. In this case, experimental set-ups of two types were used. The first type of set-up was characterized by large sizes of helical tubes, whose maximum oval size was equal to 36-52 mm. As a result, the probes inserted into the bundle at the airflow exit (Fig. 2.4) enabled studying the velocity and temperature fields under nearly isothermal conditions. The second type of experimental set-up was distinguished by the fact that on this set-up the experiments were made with heated helical tubes. Tubes were made of corrosion-resistant steel, the tube wall thickness varied from 0.2 to 0.5 mm, and the maximum size of the oval was equal to ≈ 12 mm (Fig. 2.5). All of the set-ups were distinguished by the design of the experimental sections, the measuring system, and the system of electric power supply to the experimental section. As for the gasdynamic section, it was essentially identical for all set-ups. The tank air, via a stop cock, filter, pressure reducer controlled by a command reducer, heater, and throttle flowmeter, entered the experimental section (Figs. 2.4 and 2.5). The air flowrate was, as a rule, measured by a critical orifice meter calibrated against a gas holder. In front of the critical orifice meter, measurements were made both of pressure, using a standard manometer, and of air temperature, utilizing a six-junction chromel-copper calibrated thermocouple. In some cases, the air flowrate was measured by ordinary measuring orifice meters on which, besides the pressure and temperature in front of the orifice meter, the pressure drop was measured. Usually, orifice meter pressure drops were measured twice, first by an inductance - or capacitance-type gauge - and then by a liquid differential manometer. Let us consider the specific features of the design of the experimental sections intended for modelling heat transfer and flow in the different set-ups.

The hydraulic resistance coefficients were investigated on the set-ups of the first and second types under adiabatic airflow conditions with a reduced velocity λ = 0.04-0.250. In this case, the thermodynamic air temperature in any cross section of the model was determined by the formula:

$$T = T^* - \frac{u^2}{2c_p} \tag{2.26}$$

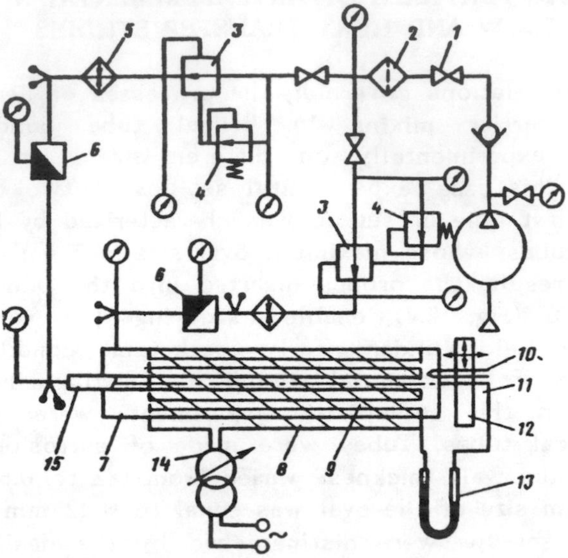

Figure 2.4 Schematic of the large-scale experimental set-up for studying the specific features of flow in a helical tube bundle: *1)* stop cock; *2)* filter; *3)* working reducer; *4)* command reducer; *5)* heater; *6)* throttle flowmeter; *7)* inlet unit; *8)* experimental section; *9)* helical tube; *10)* thermocouple; *11)* total pressure gauge or hot-wire anemometer; *12)* coordinate mechanism; *13)* differential manometer; *14)* welding transformer; *15)* movable tube.

where T^* is the stagnation temperature measured in the receiver in front of the experimental section.

The experimental section was a hexahedral steel Kh18N10T shell (Fig. 2.5). The holes, whose diameters were approximately 10 times less than the equivalent diameter, were drilled in the tube shell. A single collector united 4 or 6 holes in each cross section. As a result, it was possible to measure the static pressure at the inlet and at the outlet of the control sections of the bundle. The pressure drops on the control sections of the bundle model were measured by liquid differential manometers and different inductance-type gauges. Indications were recorded with air flowrate being increased and decreased. Geometrical sizes of the experimental sections were determined by direct measurements of the helical tube and shell sizes as well as by weighing the tube bundles.

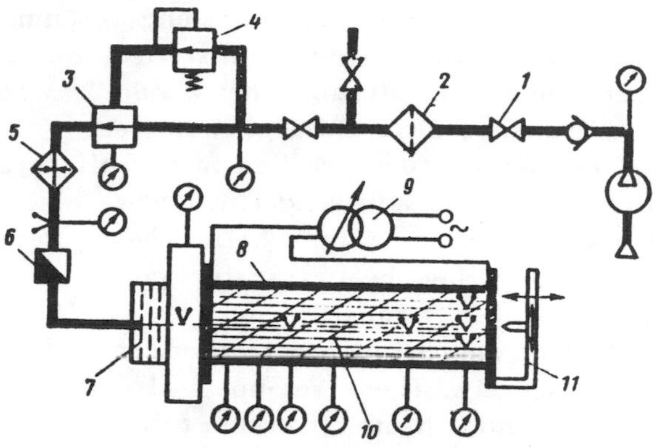

Figure 2.5 Schematic of the experimental set-up for studying heat transfer and fluid mixing with axisymmetric and asymmetric nonuniformities of heat release: *1)* stop cock; *2)* filter; *3)* working reducer; *4)* command reducer; *5)* heater; *6)* throttle flowmeter; *7)* inlet unit; *8)* experimental section; *9)* welding transformer; *10)* helical tube bundle; *11)* coordinate mechanism.

Velocity and temperature fields were studied on a bundle with axisymmetric air efflux (Fig. 2.4). On this bundle the transport flow properties were studied by the method of diffusion from a point source. The laws of jet spreading in a helical tube bundle were also examined. Therefore, the gasdynamic section of the set-up was provided with two independent heat carrier supply lines. In this case, the movable tube, via which an indicator gas was supplied, was slipped over the central helical tube of a bundle. This allowed the measurement of velocity and temperature fields in the outlet cross section of a bundle at different distances from a movable diffusion source. A bundle of 37 helical tubes was mounted into a hexahedral body, whose walls were equipped with segment inserts to support the flow cross section in the peripheral regions of a bundle. The bulk of air was supplied to the chamber ahead of the experimental section (Fig. 2.4), then it passed through a system of levelling grids and found its way into the collector ahead of a bundle. The tubes were freely mounted in the body and were fixed along the length by means of the rods relative to the inlet grid. The indicator gas parameters might either differ or be identical with those of the main flow.

Velocity and temperature distributions were measured by a total pressure tube 1.2 mm in diameter and 0.1 mm thick and by a chromel-alumel thermocouple installed on a moveable coordinate mechanism. The dynamic head, $\rho u^2/2$, was measured by an inclined U-shaped differential manometer.

Velocity and temperature profiles in a wall layer, as well as the coefficient ξ in the adiabatic flow, were also studied on the experimental set-ups (Fig. 2.5). These set-ups were intended for investigating heat transfer coefficient ξ in the nonisothermal flow at q = const(F) and q = var(F) and also the interchannel flow mixing using the method of diffusion from a system of linear sources. These set-ups had a different number of helical tubes, could heat different tube groups to study cross mixing, and had helical tube bundles of variable length. The main characteristics of these set-ups and the experimental procedures were the same.

Let us consider an experimental section with free air exhaust (Fig. 2.5) typical of all the set-ups of this type. The size chosen for this bundle enabled making measurements of velocity and temperature fields using sensors mounted on the coordinate mechanism and inserted on the side of the air exhaust, into the intertube space of a bundle. These sizes provided satisfactory parametric variation with respect to air flowrate and flow velocities at the set-up exit. The thickness of the helical tube wall was 0.2 mm. In each tube, copper current-conducting wires were soldered in the outlet cross section and then connected to a copper outlet flange. Copper pipes were soldered in the inlet cross sections via the side holes out of which the thermocouple wires could be brought (the thermocouples were welded, on the inside, to the helical tube walls). In the copper pipes, wire ends were built-in, via which electric current was supplied to the helical tubes. The wires were brought out through the pressure seals of the inlet unit and were connected with external current-carrying parts, or buses. The thermocouple wires were also brought out through the pressure seals of the inlet unit. The cold junction of the thermocouples had a temperature equal to 0°C. Thermocouples measuring the tube wall temperature were located inside the tubes in four radial directions of a bundle and in five cross sections along the tube.

For static pressure take-off, three 0.5 mm diameter holes were made in each of seven bundle cross sections in the tube shell and were connected by means of a manifold. The tube shell on the inside was coated with aluminum oxide, which is a thermo-stable electric insulation ($T \leq 700$ K).

The input unit of the experimental section, to which air was supplied via a 20 mm diameter tube, promoted a uniform velocity field at the tube bundle inlet by using a special head that changed the flow direction from horizontal to radial and a system of five levelling grids, the first at a distance of 30 mm from the inlet and the remainder at intervals of 15 mm.

The tubes were heated by a.c. current from a transformer controlled by an autotransformer. The heater air was preliminarily heated by a.c. current from a transformer. During the experiment, measurements were made of the following quantities: air flowrate, temperature and pressure at the inlet and outlet of the experimental section, current supplied to the bundle, wall temperature of the helical tubes, and bundle pressure distribution. Pressure drops on the control sections were measured by liquid differential manometers. The temperature and velocity head at each point of a design bundle cross section at a distance of 25 mm from the bundle outlet upstream were determined by the sensors located on the coordinate mechanism.

To improve the reliability of the data, the flowmeter and thermocouples were calibrated, individual parameters were measured several times using different instruments as well as the experimental set-up, and the measuring system was operated using adopted procedures. The measuring system used made it possible to determine the hydraulic resistance coefficient within a limiting error of ±9%, and the Nusselt number within a limiting relative error of ±7%. A heat balance was performed with an accuracy of ≈ 4%.

On these experimental sections, uniform axisymmetric air entry into a helical tube bundle and free efflux were provided to study local flow characteristics (Fig. 2.5). Therefore, it proved possible to investigate the heat transfer on hydrodynamic and thermal stabilization lengths as well as the velocity and temperature fields. An increase in the number of helical tubes from 19 to 37 and 127 improved the heat balance for the central zone of a bundle by creating flow conditions approaching those in an infinite grid.

A helical tube bundle of the set-up (Fig. 2.5) for studying gas mixing promoted heating of individual groups of helical tubes. For this, the helical tubes were electrically insulated by a thermo-stable insulating varnish.

To measure the local flow characteristics in the cells of a bundle cross section, the heat carrier efflux from the experimental section is free and the bundle outlet is equipped with the coordinate mechanism, on which total pressure gauges and thermocouples are located. These devices make possible simultaneous comprehensive studies of heat transfer and flow in helical tube bundles.

Let us emphasize the principal features of an experimental set-up composed of 127 tubes [17].

This set-up is an open-type tube, to which air was supplied by means of a turbine compressor. Air flowrates were measured by double diaphragms and controlled by remotely controlled valves mounted on the main and bypass lines. Air was admitted into the bundle from the bottom upwards via a diffusor, a chamber where the flow is smoothed by grids, and via an inlet Vitoshinsky-profile nozzle. Having passed through the bundle, the air was exhausted into the atmosphere.

A bundle of 37 helical tubes was heated axisymmetrically by a d.c. current generator. Its power was controlled by changing current in the field circuit of the generator. A special feedback electronic device was used to stabilize the voltage. The current intensity was measured by a shunt voltage drop.

Tube wall and air temperatures were measured by chromel-alumel thermocouples connected with automatic systems for data acquisition and recording. The thermocouples measuring wall temperatures were welded to the inner surface of the helical tubes and the thermocouple wires were brought out via the lower part of the bundle and via the inlet channel wall to a measuring device. Pitot tubes were used to measure flow velocities in the bundle outlet cross section.

Temperatures and velocities were measured in the bundle cross section at a distance of 25 mm upstream from the bundle outlet cross section. A heat carrier temperature sensor and a flow velocity meter were mounted on a special coordinate mechanism. The above devices made it possible to study heat transfer at nonuniform heat supplies over the radius of a helical tube bundle.

A large-scale set-up of the first type was used to determine the averaged velocity vector components. In these studies, a hot-wire anemometer was also mounted on the coordinate mechanism. To measure all velocity components it was necessary to locate the indications of this anemometer at different positions. Therefore, the measuring time at one point increased. Recording of the anemometer output signal and other measured quantities was automated to improve accuracy. In this case, the anemometer signals were transformed into a digital form and were transferred to the information data storage of a measuring-computer complex.

In all the above-mentioned set-ups intended for studying the flow structure and interchannel mixing as well as the hydraulic resistance and heat transfer, air was axially supplied to the bundle and, thus, the heat transfer was mainly investigated on the central tube of the bundle. Unlike these set-ups, another device was designed to explore the mean heat transfer performance in a bundle. To achieve more closley the flow conditions typical of heat exchangers, the flow at the bundle inlet and outlet in this device was at crossflow conditions. A schematic of this device is shown in Fig. 2.6.

Air from the compressor-filled tanks enters through the coil of an auxiliary heater, a reducer, an oil filter, a diaphragm - flow meter, and then enters the experimental section. The air flowrate is regulated by a reducer. Air from the experimental section, via a coil-cooler and valve that regulates the flow cross section at the bundle outlet, is exhausted into the atmosphere. Air is supplied in the transverse direction to the intertube space by way of equidistant holes of the upper casing. Air is also removed from the experimental section in the transverse direction through two outlet connections.

The experimental section is located vertically. The heat exchanger is a bundle of 19 helical tubes placed into a hexahedral casing. Comparison of the data obtained on this set-up with those for the bundles with a great number of tubes indicated that the effect of the casing wall on heat transfer and hydraulic resistance for a 19-tube bundle is insignificant. This allows a successful modeling of the tube bundles on this bundle at relatively small power inputs to the experimental set-up.

Experiments were made on three bundles of helical tubes having different twisting pitches. Steel 1Kh18N10 helical tubes 0.4 mm thick made of round tubes 6 mm in outer diameter were simultaneously twisted and pressed out on the turning lathe using a special instrument. The twisting pitch of the bundles was s=90, 60, and 30 mm in relative quantities s/d = 12.45, 8.30, and 4.15, respectively (d = 7.24 mm).

In all tubes, the length of a twisted section was 750 mm. Both their untwisted cylindrical ends were inserted into the holes of the tube plates (Fig. 2.6). The tubes were directly heated by a.c. low-voltage current supplied via copper tube

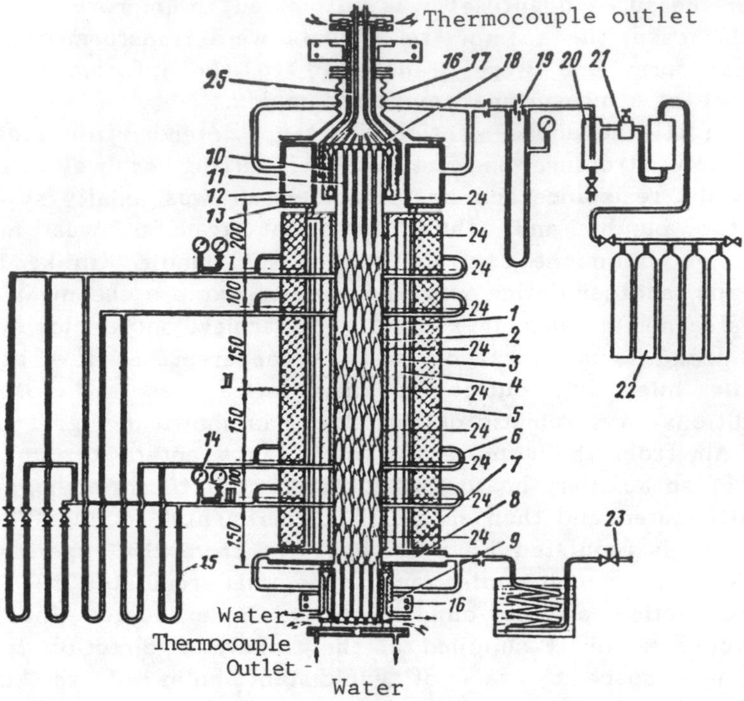

Figure 2.6 Schematic of the experimental set-up for studying mean heat transfer and hydraulic resistance with uniform heat release: *1)* helical tubes; *2)* insulation plates; *3)* casing; *4)* screen; *5)* tube shell; *6)* pressure holes; *7)* thermocouple; *8)* air exhaust; *9)* lower tube plate; *10)* upper tube plate; *11)* tin; *12)* thermocouple; *13)* upper body; *14)* manometer; *15)* differential manometer; *16)* current-carrying busbar; *17)* siphon; *18)* thermocouple electrodes; *19)* diaphragm for measuring air flowrate; *20)* filter; *21)* reducer; *22)* tanks; *23* controlling valve; *24)* voltage take-off; *25)* current supply

plates. After 19 tubes were arranged in a bundle and were assembled with the tube plates, the inner cavities of the latter were filled with melted tin (Fig. 2.6) to a depth of 70 mm. This provided reliable electric contact between the tubes and tube plates, which was important for attaining a uniform current distribution over the tubes. The tube plates were cooled with water to prevent further tin melting during the experiment.

Before the assembly was completed, the electric resistance of each tube was measured in preliminary experiments. The maximum nonuniformity of the current distribution over the tubes of the bundle due to mutual inductance did not exceed 0.45%. The ohmic resistance of the tubes as a function of temperature was consistent with known data for 1Kh18N10 steel.

In assembly, the tubes were oriented symmetrically with one another. Once this geometrical assembly was achieved, it was possible to systematize and analyze the experimental data (in particular, with respect to temperature nonuniformities).

The hot junctions of the chromel-alumel thermocouples were embedded flush with the surface of the tube walls. Thermoelectrodes of 0.2 mm diameter thermocouples having thermo-stable glass-fiber insulation were brought out via the inner tube cavities and sealing rubber spacers of the upper and lower flanges of the experimental sections (Fig. 2.6).

A tube bundle was placed into a steel hexahedral casing consisting of two halves which were sealed by paronite spacers. The inner casing cavity was lined with mica glass-ceramic electrically insulating plates. The sizes of the inner casing cavity together with the mica glass ceramic plates promoted closed packing (relative contact) of the bundle tubes.

The size of the cross section of the tubes and its area, as well as areas of the cross section of the body in assembly and the flow cross section of a bundle, were determined from linear measurements and from the measured volume of water fed into the experimental section. The main geometrical sizes of an experimental section are listed in Table 2.1.

Two screens and a thermally insulating casing were mounted to decrease heat losses on the outside of the tube shell.

The upper tube plate could move in the longitudinal direction relative to the casing to compensate for the difference in thermal expansions of the tube bundle and the casing. A current-conducting cylinder was screwed on the upper tube plate. Electric circuit elementys (busbars, tube plates, test bundle, current supply) were thoroughly insulated from the remaining components of the experimental section (Fig. 2.6).

Electric power supplied to the experimental section was controlled by an autotransformer. Maximum electric power amounted to 90 kW and voltage on this section varied from 4 to 20 V; current varied from 400 to 2500 a.

Parameter	Notations	Dimension	s/d 12.45	8.30	4.15
Maximum tube profile size	d	mm	7.23	7.25	7.24
Minimum profile size	Δ	mm	4.15	4.13	4.51
Cross-sectional area of one tube	f_{area}	mm^2	25.15	25.11	25.9
Perimeter of tube cross section	Π_{tube}	mm	18.85	18.85	19
Distance between opposite sides of casing	H	mm	33.9	33.4	33.9
Area of the flow cross section of casing	F_k	mm^2	991.7	961	991.7
Area of the flow cross section of bundle	F	mm^2	514	484	500
Perimeter of the casing cross section	Π_k	mm	116.1	116.1	116.1
Total wetted wall perimeter	Π	mm	476	475.5	477.5
Equivalent diameter with respect to the total wetted wall perimeter	d_{eq}	m	4.32	4.08	4.19
Equivalent diameter of a single central cell	d_{eq}	mm	4.32	4.32	4.11

The air flowrate varied from 6 to 120 g/s and was measured by diaphragms placed in 12 and 20 mm diameter pipelines. Coefficients of air flowrates through these diaphragms were determined by special calibrations using a measuring tank.

The hydraulic resistance of a bundle was studied on a section l_0 = 500 mm at distances of 200 and 150 mm from its inlet and outlet, respectively.

Static pressure on the section l_0 was measured in four cross sections at distances of 200, 400, 600, and 700 mm, or x/d_{eq} = 48.6, 72.4, 143, and 167, respectively, from the upper tube plate. Pressure holes 0.8 mm in diameter were on the opposite casing sides (two in each cross section). Measurements were made of pressure drops between these holes and on the section l_0.

Wall temperatures of the helical tubes were measured in 9 cross sections along the bundle length. Heat transfer was studied on the central bundle section l_0 = 500 mm. In the main cross sections *I, II,* and *III* of the central part of the bundle where the heat transfer coefficient was determined, all 19 tubes were provided with thermocouples. These cross sections were located at distances x/d_{eq} = 48.6, 108, and 167 from the bundle inlet. In cross section *III*, three tubes had two hot junctions embedded flush with the flat and oval parts of the tube, which allowed an estimation of the nonuniformity of the temperature distribution over the tube perimeter. In the remaining tubes, the hot junctions were located in the middle of the flat part of the tube. The central tube had 9 thermocouples placed at distances x/d_{eq} = 7.12, 13.1, 24.9, 48.6, 78.4, 108, 137, 167, and 196 from the bundle inlet.

The temperatures of the screens and thermal insulating casing were measured to determine heat losses in 4 cross sections along the tube bundle length. Measurements were also made of the inlet air temperature at three points, outlet air temperature at two points, current supplied to the exeprimental section, and voltage drop on 9 sections of a bundle (using voltage take-off 24 in Fig. 2.6).

On this experimental set-up, the mean heat transfer coefficients on the central section for all tubes of a bundle as well as the local heat transfer coefficients (mean for a given cross section) for cross sections *I, II,* and *III* were determined. As shown below, the choice of distances of these cross sections from the tube bundle inlet and outlet was stipulated by a necessity to eliminate the effect of air cross-supply and removal on the heat transfer and hydraulic resistance.

Heat flux density from a tube bundle to the air was determined in terms of the measured intensity of current I passing through a bundle and in terms of an electric temperature-dependent resistance of the bundle:

$$q = \frac{I^2 R(\bar{T}_w)}{n\Pi_{tube}} = \frac{I^2 \rho(\bar{T}_w)}{nS_{tube}\Pi_{tube}} \tag{2.27}$$

where $R(\bar{T}_w)$ is the electric resistance of one meter of a bundle, $\rho(\bar{T}_w)$ the specific resistance of a tube bundle, n the number of tubes, Π_{tube} the perimeter of a tube cross section, S_{tube} the cross-sectional area of the tube walls, and $\bar{T}_w$ the bundle wall temperature in a given cross section determined for all tubes of a bundle.

The local heat transfer coefficient along the length and mean over the cross section was

$$\alpha = \frac{q}{\bar{T}_w - \bar{T}_f} \tag{2.28}$$

where T_f is the mean mass flow temperature in a given cross section and is determined through the flow enthalpy increment from the inlet cross section of a bundle

$$\bar{T}_{f\,l+1} = \bar{T}_{f\,l} + \frac{\Delta Q_{el}(i,\ i+1) - \Delta Q_{h.loss}(i,\ i+1)}{Gc_p\ (i,\ i+1)} \tag{2.29}$$

where i and i+1 are the numbers of cross sections, ΔQ_{el} the heat release on the section between the cross sections i and i+1, $\Delta_{h.loss}$ the heat losses through the casing between these cross sections, c_p the mean heat capacity over the temperature range $(T_{fi} \ldots T_{f\,i+1})$, and G the air flowrate. This temperature was close to the one determined, assuming that the inlet and outlet flow temperatures changed linearly.

The mean heat transfer coefficient for the section l_0 was determined as:

$$\bar{\alpha} = \frac{\int_{x_1}^{x_1+l_0} q(x)\,dx}{\int_{x_1}^{x_1+l_0} (\bar{T}_w - \bar{T}_f)\,dx} \tag{2.30}$$

where x_1 is the coordinate of the entrance length of a section. The mean heat transfer coefficient for 7 central tubes of a bundle was determined to estimate the effect of the casing wall on heat transfer. In this case, the value of T_f was taken the same as the entire bundle in a given cross section. Thus, the estimate of a heat transfer difference between 7 central and all 19 tubes of a bundle was a maximum. In determining the heat flux density for the 7 central tubes, an account was taken of the difference between their temperature and the temperature of the remaining bundle tubes.

The hydraulic resistance coefficient, ξ, of a bundle for the section between the cross sections i and i+1 was determined by the equation:

$$\Delta p_{i,i+1} = \xi \frac{\overline{\rho u^2}}{2} \frac{\Delta x}{d_{eq}} + \left(\rho_{i+1} u_{i+1}^2 - \rho_i u_i^2\right) \tag{2.31}$$

where $\Delta p_{i,i+1}$ is the measured static pressure drop; ρ and u are the mean density and velocity for the test section; $\rho_i u_i$ and $\rho_{i+1}\ u_{i+1}$ are the same quantities at the inlet and outlet of the considered section.

Measurement of the temperature of individual bundle tubes enabled determining the nonuniformity of its distribution in each cross section. Temperature nonuniformities were not specially produced and were caused by different engineering reasons as well as by the varying flow temperature over a bundle cross section due to radial heat losses.

As all the analyzed bundles with different twisting pitches were under almost the same conditions, it may be assumed that the determination of temperature nonuniformity allowed a qualitative estimation of the effect of the tube twisting pitch on leveling of temperature nonuniformities in a tube bundle. Data on 7 central tube bundles were processed, since the temperatures of the peripheral tubes were affected by the casing. The nonuniformity was estimated by a rms deviation

$$\delta T_w = \frac{\sqrt{\dfrac{\sum_1^n (T_{wi} - \bar{T}_{cf})^2}{n}}}{\bar{T}_{cf} - \bar{T}_f} \tag{2.32}$$

where T_{wi} is the wall temperature of one of the central tubes (in cross sections *I*, *II*, or *III*), $\bar{T}_{c.f.}$ is the mean temperature of the central tubes, and $(\bar{T}_{c.f.} - \bar{T}_f)$ is the temperature head in a given cross section, n=7.

This experimental set-up was used to determine hydraulic resistance coefficients with a limiting error of ±9% and the Nusselt number within ±10%. The difference in the heat balance was ±11%.

CHAPTER
THREE

THE STRUCTURE OF LONGITUDINAL TURBULENT FLOW IN HELICAL TUBE BUNDLES

3.1 THEORETICAL MODELS FOR THE FLOW IN A HELICAL TUBE BUNDLE

Different theoretical flow models may be used to solve the problems of heat transfer and mixing in helical tube bundles and to calculate the temperature and velocity fields. The model for "vortex" flow in a bundle is one of the flow models which was adopted to obtain the Fr_M number in § 1.4. According to this model, the flow in helical channels of tubes is swirled as a solid

$$u_\tau r^{-1} = \text{const} \tag{3.1}$$

As shown below, this law for the swirling flow is mainly satisfied only in the external part of a wall layer of the tubes. In the flow core, tangential velocity components are determined by interacting "vortices" induced by the tubes. Thus, on a line connecting two adjacent tubes, the velocities u_τ in the flow core decrease to zero at the conventional boundary of two contacting helical channels and then reverse direction since the flow in these channels has an opposite swirl. In the flow core of the peripheral part of a bundle where there is no interaction of different-direction "vortices," the velocity u_τ changes according to the constant circulation law, $u_\tau \cdot r$=const. This flow model, although it is physically grounded and may be employed to generalize and to analyze experimental data, does not allow thermal and hydraulic calculations to be made with regard to interchannel mixing.

A model, in which the flow in a real tube bundle may be replaced by one in a system of parallel channels, can be used as a flow model that allows thermal and hydraulic calculations to be made with respect to the mean flow parameters. However, this model is not physically grounded and does not allow for heat exchange between individual cells of a bundle.

It is obvious that preference should be given to a flow model in which the bundle is replaced by a porous mass. The diameter of this bundle is equal to that of a bundle in which a homogenized medium is flowing, i.e., the flow of the heat carrier with distributed sources of volumetric energy release q_V and hydraulic resistance $\xi\ \rho u^2/2$, whose intensities vary with respect to the bundle radius and azimuth.

Replacement of the flow in a real tube bundle by a homogenized medium is an engineering procedure. The validity of its application for the calculation of temperature and velocity fields of the heat carrier must be checked by experiment. Although a strict mathematical statement of the problem in this case is inconvenient, this flow model can be described by a system of equations that can be solved and which allows thermal and hydraulic calculation to be made with regard to interchannel mixing and the velocity and temperature fields of the heat carrier. In this case, account is taken of the fact that the swirling motion of the spiral flow must substantially expand the region of the flow and decrease the wall layer thickness. Then, after the wall layer displacement thickness δ^* is determined and a material layer equal to the thickness δ^* is built up on the tube walls, it is possible to consider, within limits, the free flow involving a homogenized medium slip, assuming that the velocity vector is parallel to the bundle axis and $\partial p/\partial r = 0$.

The validity of the assumption that $\rho \approx \text{const}(r)$ in longitudinal flow past helical tube bundles has been further verified by experiment. This is due to the fact that the thickness of the wall layer on helical tubes is substantially less than the curvature radius of the helical channel wall ($R \gg \delta$) and that the curvature of the helical surface of the tubes does not undergo sharp changes.

When the stream in the flow core is considered with a uniform heat supply with respect to the radius of a helical tube bundle, the equation of motion (1.6) does not account for all of the viscosity-dependent terms, since in this flow region the velocity gradient normal to the wall is practically absent and the Reynolds numbers are large:

$$\frac{dp}{dx} = -\bar{\rho}\bar{u}\frac{d\bar{u}}{dx} \tag{3.2}$$

The bars above T, u, and ρ in Eq. (3.2) and in other equations means that these parameters refer to the flow core. However, Eq. (3.2) must take into consideration the distributed sources of hydrualic resistance $\xi\ \bar{\rho}\ \overline{a^2}/2d_{eq}$ in accordance with the homogenized flow model. Therefore, instead of expression (3.2) we have:

$$\frac{dp}{dx} = -\bar{\rho}\bar{u}\frac{d\bar{u}}{dx} - \xi\frac{\bar{\rho}\bar{u}^2}{2d_{eq}} \tag{3.3}$$

In this case, the flow outside the boundary layer and its boundary can be determined from calculation in terms of mean parameters and from balance calculations: $\bar{u} = \bar{u}(x)$ = const (r), $\bar{T} = \bar{T}(x)$ = const (r), $\bar{p} = \bar{p}(x)$ = const (r). It may be shown that the equation of motion in the form (3.3) can be obtained from Eq. (1.6) in cylindrical coordinates for the tube bundle flow:

$$\rho u\frac{d\bar{u}}{dx} = -\frac{dp}{dx} + \frac{1}{r}\frac{\partial}{\partial r}(r\tau) \tag{3.4}$$

Indeed, if both sides of the equation are multiplied through by r and integrated with respect to the radius, i.e.,

$$\bar{\rho}\bar{u}\frac{d\bar{u}}{dx}\int_0^{r_{cell}} r\,dr = -\frac{dp}{dx}\int_0^{r_{cell}} r\,dr - \Big|_0^{r_{cell}} r\tau$$

assuming that $d\bar{u}/dx$ and dp/dx do not depend on the bundle cell radius and that τ changes linearly with respect to r (on the cell axis τ = 0), we obtain

$$\bar{\rho}\bar{u}\frac{d\bar{u}}{dx} = -\frac{dp}{dx} - \frac{2}{r_{cell}}\tau_w \tag{3.5}$$

where τ_w = (1/8) $\xi\bar{\rho}\bar{u}^2$ is the shear stress on the helical tube wall. Then, considering that in a helical oval tube bundle $d_{eq} \approx 2r_{cell}$ (r_{cell} is the bundle cell radius) and substituting into (3.5), instead of τ_w, its expression in terms of ξ, we obtain the equation in the form (3.3).

The equation of motion for uniform heat supply in the bundle cross section should also incorporate diffusional terms, since a nonuniform temperature field of the heat carrier also promotes velocity nonuniformities in the bundle cross section. Inclusion of these terms into Eq. (3.3) should not affect the integral characteristics of the bundle, i.e., in integrating the equation of motion with respect to the bundle radius r and the azimuthal coordinate φ, the diffusional terms must be omitted. Let us write the motion equation with the diffusional terms in cylindrical coordinates:

$$\bar{\rho}\bar{u}\frac{\partial \bar{u}}{\partial x} = -\frac{dp}{dx} + \frac{1}{r}\frac{\partial}{\partial r}\left(r\nu_{ef}\frac{\partial \bar{u}}{\partial r}\right) + \frac{1}{r^2}\frac{\partial}{\partial \varphi}\left(\nu_{ef}\frac{\partial \bar{u}}{\partial \varphi}\right) - \xi\frac{\bar{\rho}\bar{u}^2}{2d_{eq}} \tag{3.6}$$

Upon integration with respect to the radius and azimuth of a bundle, this equation is reduced to the equation of motion in the form:

$$(\bar{\rho}\bar{u})_{mean}\frac{\partial \bar{u}_{mean}}{\partial x} = \frac{dp}{dx} - \xi\frac{\bar{\rho}_{mean}\bar{u}^2_{mean}}{2d_{eq}} \tag{3.7}$$

where the parameters entering it are the mean parameters over the bundle cross section, and

$$\int_0^{2\pi}\int_{-r_c}^{r_c}\frac{1}{r}\frac{\partial}{\partial r}\left(r\nu_{ef}\frac{\partial \bar{u}}{\partial r}\right)drd\varphi + \int_0^{2\pi}\int_{-r_c}^{r_c}\frac{1}{r^2}\frac{\partial}{\partial \varphi}\left(\nu_{ef}\frac{\partial \bar{u}}{\partial \varphi}\right)drd\varphi = 0$$

where r_c is the radius of the heat exchanger casing.

For the axisymmetric problem, the system of equations for flow of a homogenized medium with distributed sources of volumetric energy release and hydraulic resistance with regard to a displacement thickness, δ^*, of a boundary layer is of the form:

$$\bar{\rho}\bar{u}\frac{\partial \bar{u}}{\partial x} = -\frac{dp}{dx} + \frac{1}{r}\frac{\partial}{\partial r}\left(r\nu_{ef}\frac{\partial \bar{u}}{\partial r}\right) - \xi\frac{\bar{\rho}\bar{u}^2}{2d_{eq}} \tag{3.8}$$

$$\bar{\rho}\bar{u}c_p\frac{\partial \bar{T}}{\partial x} = q_v\frac{1-m}{m} + \frac{1}{r}\frac{\partial}{\partial r}\left(r\lambda_{ef}\frac{\partial \bar{T}}{\partial r}\right) \tag{3.9}$$

$$G = 2\pi m\int_0^{r_c}\bar{\rho}\bar{u}r\,dr \tag{3.10}$$

$$p=\bar{\rho}R\bar{T} \tag{3.11}$$

with boundary conditions

$$\bar{u}(0,\ r)=\bar{u}_{in},\ \bar{T}(0,\ r)=\bar{T}_{in},\ p(0,\ x)=p_{in} \tag{3.12}$$

$$\left.\frac{\partial\bar{u}(x,\ r)}{\partial r}\right|_{r=r_c}=0,\ \left.\frac{\partial\bar{T}(x,r)}{\partial r}\right|_{r=r_c}=0 \tag{3.13}$$

$$\left.\frac{\partial\bar{u}(x,r)}{\partial r}\right|_{r=0}=0,\ \left.\frac{\partial\bar{T}(x,\ r)}{\partial r}\right|_{r=0}=0 \tag{3.14}$$

This system of equations neglects the heat and momentum transfer due to molecular diffusion, the heat release with flow kinetic energy dissipation, and longitudinal turbulent diffusion.

System (3.8)-(3.11) has no convective terms with transverse velocity components $\bar{\rho}\bar{v}\ (\partial\bar{u}/\partial r)$, $c_p\bar{v}\bar{\rho}\ (\partial\bar{T}/\partial r)$, etc. since it is assumed that the velocity vector is parallel to the axis of the tube bundle. Actually, the transverse velocity components in a helical tube bundle, although less by an order of magnitude than the longitudinal component, may exert a pronounced effect on the improvement of heat and mass transfer processes. It is assumed that this effect is allowed for in the experimentally determined coefficients ν_{ef} and λ_{ef} in the model helical tube bundles. It should be noted that the majority of the assumptions adopted are generally accepted in the case of free flow with a homogenized medium slip, the homogenization effect being taken into account by including a multiplier $(1-m)m$ into the energy equation.

The coefficients ν_{ef}, λ_{ef}, and ξ can be found experimentally. In this case, according to (1.24), an assumption may be made:

$$\mathrm{Pr}_{\tau}=\frac{c_p\nu_{ef}}{\lambda_{ef}}=1 \tag{3.15}$$

It is also assumed that the Lewis number is

$$\mathrm{Le}_{\tau}=\frac{\rho c_p D_t}{\lambda_{ef}}=1 \tag{3.16}$$

Then, the problem of the closure of system (3.8)-(3.11) may be reduced to a determination of the effective diffusion coefficient D_t from experiment or in a dimensionless form:

$$k = \frac{D_t}{\bar{u} d_{eq}} \tag{3.17}$$

The coefficient D_t allows for the action of all transfer mechanisms typical of helical tube bundles. These are turbulent diffusion, convective transfer due to the vortex motion in bundle cells, ordered transfer by helical channels of the tubes, and secondary flow circulations. The velocity and temperature fields calculated for such flows are external relative to the wall layer. In this case, a velocity $\bar{u}$ and temperature $\bar{T}$ are determined at the external boundary of the wall layer. Knowing $\bar{u}$ and $\bar{T}$, boundary layer methods may be used to calculate the velocity and temperature fields in the wall layer. In the wall layer, the flow velocity changes from 0 to $\bar{u}$, while the heat carrier temperature changes from T_w to $\bar{T}$.

Boundary layer equations (1.10)-(1.15) for steady (mean) liquid flow in the wall layer of helical tubes, when heat carrier dissipation is neglected, will be of the form:

$$\frac{\partial(\rho u)}{\partial x} + \frac{\partial(\rho v)}{\partial y} = 0 \tag{3.18}$$

$$\rho u \frac{\partial u}{\partial x} + \rho v \frac{\partial u}{\partial y} = -\frac{dp}{dx} - \frac{d\tau}{dy} \tag{3.19}$$

$$\rho u c_p \frac{\partial T}{\partial x} + \rho v c_p \frac{\partial T}{\partial y} = \frac{\partial q}{\partial y} \tag{3.20}$$

Equations (3.18)-(3.20) for a flat plate boundary layer may be used for calculating a uniform liquid whose physical properties depend on pressure and temperature, in steady subsonic turbulent flow. Equations (3.18)-(3.20) must be supplemented with expressions (1.16) and

$$q = -\rho c_p \overline{v'T'} + \lambda \frac{\partial T}{\partial y} = \rho c_p \left(\frac{\nu}{\mathrm{Pr}} + \frac{\nu_\tau}{\mathrm{Pr}_\tau} \right) \frac{\partial T}{\partial y} \tag{3.21}$$

$$\rho = \rho(p, T),\ c_p = c_p(p, T),\ \mu = \mu(p, T),\ \lambda = \lambda(p, T) \tag{3.22}$$

where x and y are the coordinate axes along the wall downstream and normal to the wall, respectively.

Semiempirical turbulence theories may be employed to close the system of equations (3.17)–(3.22). Also, to a first approximation the following relations may be used:

$$\frac{\nu_T}{\nu}=0.4\left\{\eta-7.15\left[\operatorname{th}\left(\frac{\eta}{7.15}\right)+\frac{1}{3}\operatorname{th}^3\left(\frac{\eta}{7.15}\right)\right]\right\} \quad (3.23)$$

$$\left.\begin{aligned}\Pr_T &\approx 1.0\\ \eta &= \frac{l\sqrt{\tau/\rho}}{0.4\nu}\end{aligned}\right\} \quad (3.24)$$

$$\left.\begin{aligned}&\frac{l}{\delta}\approx 0.14-0.08\left(1-\frac{y}{\delta}\right)^2-0.06\left(1-\frac{y}{\delta}\right)^4\\ &\text{or at small } y/\delta;\ l\approx 0.4y\end{aligned}\right\} \quad (3.25)$$

Thus, the boundary layer flow on helical tubes may be governed by the system of equations for a turbulent boundary layer of a fluid with variable physical properties. This system of equations may be used to calculate velocity and temperature fields in the wall layer with the boundary conditions:

$$y=0,\ u=0,\ T=T \quad (3.26)$$

$$y=\delta,\ u=\bar{u};\ T=\bar{T} \quad (3.27)$$

assuming that the flow parameters at the external boundary of a wall layer are predetermined. However, in practice, we may confine ourselves to solving the integral momentum and energy equations for a boundary layer to determine the wall temperatures along each helical tube and to calculate the development of the boundary layer along the wall in the flow at $0 < y < \delta$ (Chapter 1).

These models considered for the helical tube bundle flow will be used in the next sections to analyze the experimental data on flow structure, heat transfer, hydraulic resistance, and interchannel mixing as well as to calculate the temperature and velocity fields in such bundles.

3.2 THE RESULTS OF EXPERIMENTAL STUDIES ON VECTOR DISTRIBUTIONS OF AVERAGE VELOCITY AND TEMPERATURE

Velocity and temperature fields were examined on the models of a 37 helical tube bundle located in a hexahedral casing. Maximum tube cross sections of d=46 and 12.2 mm with relative twisting pitches $s/d \approx 13$, 19.6, and 24.6 were considered. The bundle lengths were l = 0.5 and 1 m. Oval and three-blade tube bundles [7] were studied. A total pressure tube (§ 2.2) was used to measure velocities, and chromel-alumel thermocouples mounted on the coordinate mechanism recorded the temperature. Experiments were made at Re = $4.5 \cdot 10^3$–10^4, Fr_M = 64–1530, porosity $m \approx$ 0.3–0.6, and velocity V = 5–40 m/s. Experimental data on the velocity fields were given in the form of Eqs. (1.57) and (1.59).

From an analysis of the velocity field it was found that a considerable region of approximately constant velocity and a relatively thin wall on the tubes (Figs. 3.1 and 3.2) are observed in a helical tube bundle. There is a difference beween the velocity field in a helical tube bundle and that in a circular tube, where on the stabilized flow length the wall layer thickness is equal to the tube radius [55], and the one in a circular tube bundle, where the wall layer occupies the entire flow region although it has a tube thickness variable over the perimeter [49]. As seen in Fig. 3.1*a* and Fig. 3.2*a*, the wall layer thickness in a helical tube bundle depends on Fr_M. The smaller Fr_M, the smaller the wall layer thickness. The wall layer thickness in a helical tube bundle is variable with respect to the perimeter. It is larger at the center of spiral tube channels and is smaller on the ends of the blades located in the through channels of the bundle (Fig. 3.2*a*). If d_{eq} is the characteristic dimension of a tube bundle, then, as follows from Figs. 3.1*a* and 3.2*a*, the velocity profile in the wall layer of helical tubes is more fully compared to the one in the equivalent tube:

$$\frac{V}{\bar{V}} = \left(\frac{2y}{d_{eq}}\right)^{1/7} \tag{3.28}$$

This may be explained by the centrifugal force impact. A similar phenomenon is also observed in the case of flow in tubes with spiral inserts [47].

With increasing Re, the velocity profile in the wall layer of helical tubes becomes more full (Fig. 3.1*a*). If a local wall layer thickness δ is taken as the characteristic dimension of the bundle, then the velocity fields in Figs. 3.1*a* and 3.2*a* may be described by a power law (Figs. 3.1*b* and 3.2*b*)

$$\frac{V}{\bar{V}} = \left(\frac{y}{\delta}\right)^{1/n} \tag{3.29}$$

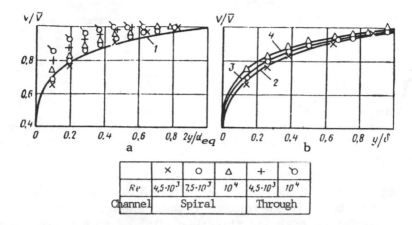

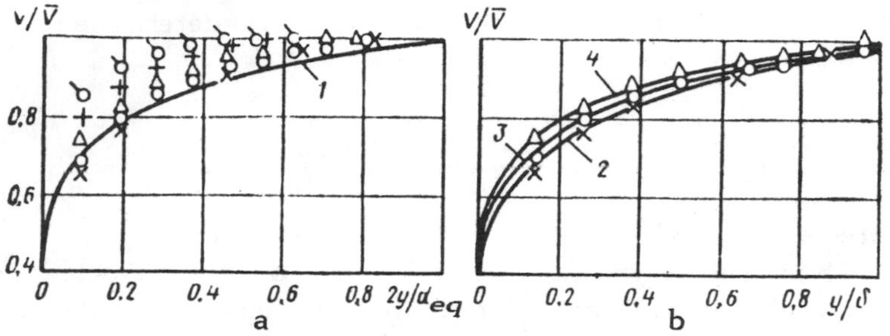

	×	○	Δ	+	ɤ
Re	$4.5\cdot10^3$	$7.5\cdot10^3$	10^4	$4.5\cdot10^3$	10^4
Channel	Spiral			Through	

Figure 3.1 Reynolds number effect on averaged velocity vector modulus distribution in a bundle of three-blade tubes with Fr_M = 1530. Equivalent diameter (*a*) and local wall layer thickness (*b*) are used as characteristic dimensions: *1*) equation (3.28); *2*, *3*, *4*) power law (3.29) and n = 3.23, 6, and 7, respectively.

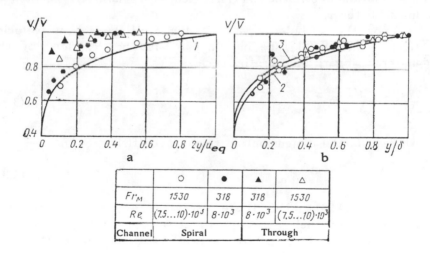

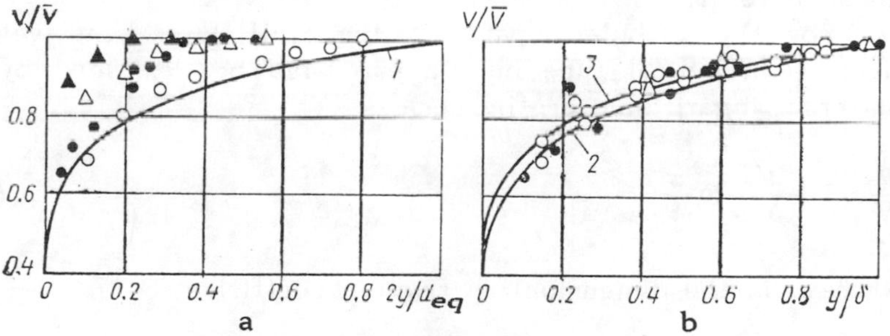

	○	●	▲	Δ
Fr_M	1530	318	318	1530
Re	$(7.5...10)\cdot10^3$	$8\cdot10^3$	$8\cdot10^3$	$(7.5...10)\cdot10^3$
Channel	Spiral		Through	

Figure 3.2 Effect of the Fr_M number on an averaged velocity vector modulus distribution in a helical tube bundle. Equivalent diameter (*a*) and local wall thickness (*b*) are used as characteristic dimensions: equation (3.28); 2,3, power law (3.29) at n = 6 and 7, respectively.

where $1/n$ depends on Re (Re = $4.5 \cdot 10^3$, n = 5.25; Re = $7.5 \cdot 10^3$, n = 6; Re = 10^4, n = 7). The quantity δ is determined by the formula:

$$\frac{\delta}{d_{eq}} = A\,\mathrm{Fr}_M^{0,32} \tag{3.30}$$

where A = 0.0349 for δ_{max} and A = 0.0156 for δ_{min}.

Thus, the Reynolds number effect on the velocity profile in the wall layer of helical tubes is qualitatively similar to that in circular tubes [55].

The experimental data for different Fr_M are correlated using relations (3.29) and (3.30) (Figs. 3.1*b* and 3.2*b*) and have proved to be universal for helical tube bundles.

Studies of oval and three-blade tube bundles have shown that approximate hydrodynamic similarity in helical tube bundles is also attained in cases with no geometrical similarity. Indeed, at a considerable change in porosity (m = 0.3–0.6) and the parameters s/d_{eq}, d/d_{eq}, h/d_{eq}, d_c/d_{eq} (d_c is the diameter of the central part of a tube) and N of the considered bundles, the velocity profiles can be described by one relation (3.29) when Fr_M is used as a characteristic number in (3.30) for δ.

For the turbulent part of the wall layer, the velocity fields in a helical tube bundle may also be described by the universal logarithmic profile [14]:

$$\frac{V}{\bar{V}} = 1 + \frac{\sqrt{\bar{\alpha}}}{0,39} \ln \frac{y}{\delta} \tag{3.31}$$

where α is the dimensionless friction coefficient

$$\alpha = \frac{\tau_w}{\rho \bar{V}^2} \tag{3.32}$$

Equations (3.29) and (3.31) are also valid for the velocity profile normal to the helical tube wall at a location where the wall layer thickness is equal to some effective wall layer thickness δ_{ef}. When δ_{ef} is taken as the characteristic dimension, the resistance law for circular tubes is satisfied. Therefore, it may be assumed that in a helical tube bundle an intercommunication exists between the velocity profile in the wall layer and the resistance law and is similar, accurate to a constant multiplier, to that existing in the flow in a circular tube or in a flat channel. For a power of 1/7 for the velocity profile it should be expected that the hydraulic resistance coefficient will be described by the relation:

$$\xi = 0.266\,\mathrm{Re}_\delta^{-0.25} \tag{3.33}$$

Actually, in steady-state flow, the fluid force equilibrium between shear stresses and pressure forces (inertia forces are absent) is expressed by the relation:

$$\left|\frac{dp}{dx}\right| = \frac{2\tau_w}{\delta\left(1 + 3.6/\mathrm{Fr}_M^{0.357}\right)^4} \tag{3.34}$$

and the dimensionless resistance coefficient ξ by the relation:

$$\left|\frac{dp}{dx}\right| = \xi\,\frac{\rho u_{mean}^2}{4}\,\frac{\left(1 + 3.6/\mathrm{Fr}_m^{0.357}\right)^{-4}}{\delta} \tag{3.35}$$

Equating (3.34) and (3.35) yields the relationship between the shear stress and the resistance coefficient:

$$\tau_w = \xi\,\frac{\rho u_{mean}^2}{8} \tag{3.36}$$

Substituting ξ from (3.33) into Eq. (3.36) and introducing the shear velocity $v_* = \sqrt{\tau_w/\rho}$, after transformation, gives:

$$\frac{u_{mean}}{v_*} = 6.99\left(\frac{\rho v_* \delta}{\mu}\right)^{1/7} \tag{3.37}$$

Let us go in (3.37) from u_{mean} to a maximum velocity $\bar{u}$, considering that in a helical tube bundle

$$u_{mean} = \bar{u}\left(1 - \frac{4\delta^*}{d_{eq}}\right) \tag{3.38}$$

For $\mathrm{Fr}_M \le 380$ it may be assumed that $u_{mean} \approx 0.965\,\bar{u}$. Then,

$$\frac{\bar{u}}{v_*} = 7.25\left(\frac{\rho v_* \delta}{\mu}\right)^{1/7} \tag{3.39}$$

Assuming that relationship (3.39) is applicable not only to the external boundary of the wall layer, we obtain:

$$\frac{u}{v_*} = 7.25\left(\frac{\rho v_* y}{\mu}\right)^{1/7} \tag{3.40}$$

For a circular tube the multiplier in (3.40) is equal to 8.74, since $u_{mean} = 0.8\ \bar{u}$ [55]. The power law of 1/7 is obtained after expression (3.40) is divided through by (3.39).

Good agreement between Eq. (3.31) and experimental data points to the fact that semiempirical turbulence theories can be used in the present flow case. Therefore, Prandtl's formula is used for the turbulent shear stress, $\tau = \rho l^2\ (du/dy)^2$, and the hypothesis is made that the "mixing" length is proportional to the distance from the wall:

$$l = \varkappa y$$

where $\varkappa$ is a dimensionless constant determined from experiment. Then, assuming $\tau = \tau_w = \rho\ v_*^2$, the asymptotic logarithmic distribution law [55] may be obtained:

$$\frac{u}{\bar{u}} = 1 + \frac{v_*}{\bar{u}\varkappa} \ln \frac{y}{\delta}$$

similar to Eq. (3.31) at $\varkappa = 0.39$.

The general use of the experimental and logarithmic velocity distributions extended to the entire region of the turbulent flow in channels may be attributed to the fact that they approximate well the real average velocity distributions. However, it should be borne in mind that experimental studies of the turbulence characteristics in a boundary layer and in a channel point to a necessity to distinguish in principle the internal and external regions of a boundary layer because of the different phenomena that occur.

In the internal region, $\approx$ (0.1–0.2) δ, the wall shear stress and viscosity (region of constant shear stress) are responsible for the flow.

In the external boundary layer section, $\approx$ (0.8–0.9) δ, the flow does not depend on the fluid viscosity. However, it is affected by the wall shear stress.

In the present investigation, by a wall layer is understood a layer δ incorporating these two flow regions. In a tube bundle, besides these flow regions, it is possible to distinguish the flow core with a practically constant velocity $\bar{u}$ = const(r).

The internal region of the wall layer, in turn, consists of a viscous sublayer, $\delta_{lam} = (0.001 - 0.1)\,\delta$, a transition region, and a turbulent layer. It should be noted that turbulent motions penetrate into the viscous sublayer up to the wall but are, by their nature, completely viscous. It is found that in each region the velocity distribution may be described by the following expressions [55]:

for the viscous sublayer

$$\frac{\bar{u}}{v_*} = \frac{v_* y}{\nu} \quad \text{or} \quad \tilde{u} = \eta;$$

for the turbulent part of the internal region

$$\tilde{u} = A \ln \eta + B$$

for the external region

$$\frac{\bar{u} - u}{v_*} = - A \ln \left(1 - \frac{y}{\delta}\right) + B^*$$

The last expression is represented in the form of the velocity defect that characterizes a velocity distribution with respect to motion at the tube (channel) center. It only deviates slightly from the logarithmic velocity distribution for the internal region of the channel flow. This enables extending this distribution for the turbulent part of the internal region to the entire turbulent flow region.

Results on temperature fields in the wall layer of helical tubes at $T_w/T_{mean} \approx 1.4$ are shown in Fig. 3.3. These excess dimensionless temperature profiles of helical tubes were measured at different distances from the bundle center. All these profiles are well described by the 1/7 power law, when δ is determined from Eq. (3.30):

$$\frac{T_w - T}{T_w - \bar{T}} = \left(\frac{y}{\delta}\right)^{1/7} \tag{3.41}$$

and those for the turbulent part of the wall layer are well described, by the logarithmic law [14]

$$\frac{T_w - T}{T_w - \bar{T}} = 1 + \frac{\alpha_m}{0.39 \sqrt{\bar{a}}} \ln \frac{y}{\delta} \tag{3.42}$$

where α_m is the dimensionless heat transfer coefficient determined by the formula

$$\alpha_m = \frac{q_w}{\rho \bar{u} c_p (T_w - \bar{T})} \tag{3.43}$$

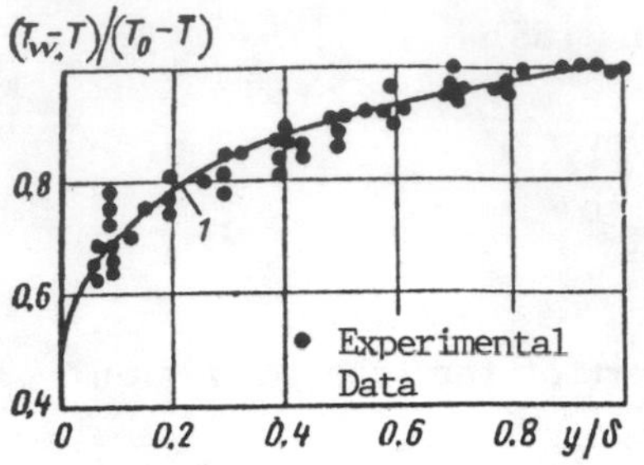

Figure 3.3 Temperature profile in the wall layer of helical tubes at Fr_M = 924 and Re = 10^4: *1*) power law (3.41)

Thus, the thickness of the thermal and velocity wall layers in a helical tube bundle have proved to be practically equal and are determined by relation (3.30). Hence, it may be considered, to a first approximation, that the assumption Pr_T = 1 in a helical tube bundle is satisfied.

Studies of the dimensionless velocity and temperature fields in bundle flow cores have shown that the observed velocity deviations in the cross section of a bundle with respect to its mean value are ≈ ±20% and are due to the vortex flow under the action of centrifugal forces in spiral channels, different flow conditions past the tubes at the locations of their contact and far from these places, as well as to accidental reasons. These deviations for temperature fields are less and amount to ±10%.

Since the rms deviations for the velocity and temperature distributions differ considerably (at Fr_M = 924, σ = 0.12 and 0.051 for the velocity and temperature fields, respectively), it may be assumed that the velocity nonuniformity in the flow core is responsible for additional flow turbulization and for smoothing the nonuniformities of the heat carrier temperature fields in the flow core and of the tube wall temperatures over the bundle cross section.

Thus, the established laws for velocity and temperature field distributions in the intertube space of swirled flow heat exchangers indicate that the entire flow region may be conventionally divided into a thin wall layer and flow core with an approximately constant velocity and temperature, i.e., these validate the model for the homogenized medium flow accepted in § 3.1. The proposed similarity number, Fr_M, considering the specific features of the flow in the intertube space of the swirled flow heat exchanger, allows generalization of the experimental data on the velocity and temperature fields.

3.3 THE RESULTS OF EXPERIMENTAL STUDY OF AVERAGE VELOCITY VECTOR COMPONENTS

Longitudinal and transverse average velocity components were studied in [8, 9, 15] by the constant-temperature anemometer method described in § 2.2 using a mini inclined-wire rotating anemometer. Nineteen oval tubes with a maximum oval size d = 36 mm formed a close-packed bundle in a hexahedral shell. Experiments were made in the range of V = 3–70 m/s, Re = $4.5 \cdot 10^3 - 1.1 \cdot 10^5$, and Fr_M = 178–1187 at pressures close to atmospheric.

It appearss that the data on the total velocity vector and the longitudinal average velocity component in a wall layer of a bundle are well described by the power law (3.29)

$$\frac{V}{V_{max}} = \left(\frac{y}{\delta}\right)^{1/n}, \quad \frac{u}{u_{max}} = \left(\frac{y}{\delta}\right)^{1/n}$$

Experimental results on the transverse velocity component distributions in the bundle cross section, where adjacent tubes do not contact, are shown in Fig. 3.4. In this cross section, measurements were made of static pressure drops $\Delta p = p_8 - p_i$ over the tube perimeter relative to the singular point of the tube profile (§ 3.4), which divides the profile into the windward and leeward sides.

As seen from Fig. 3.4, the direction of the velocity component $\bar{v}$ in the spiral tube channel is similar to that of the helical tube twisting (from a windward to the leeward side). At the conventional boundary dividing a cell into two halves, the velocity $\bar{v}$ vanishes due to the interaction of the opposite flows in the spiral channels of adjacent tubes. The relative velocity $\bar{v}/\bar{u}$ decreases with increasing Re. The velocity component $\bar{w}$ in the region of the windward side of the tube profile and in its middle is directed from the wall and in the region of the leeward side, to the wall of the tube profile (Fig. 3.4). This causes a fluid exchange bewteen the wall layer and the flow core in a cell and is one of the reasons for enhancing the fluid transport and heat transfer in a helical tube bundle (Chapters 4 and 5).

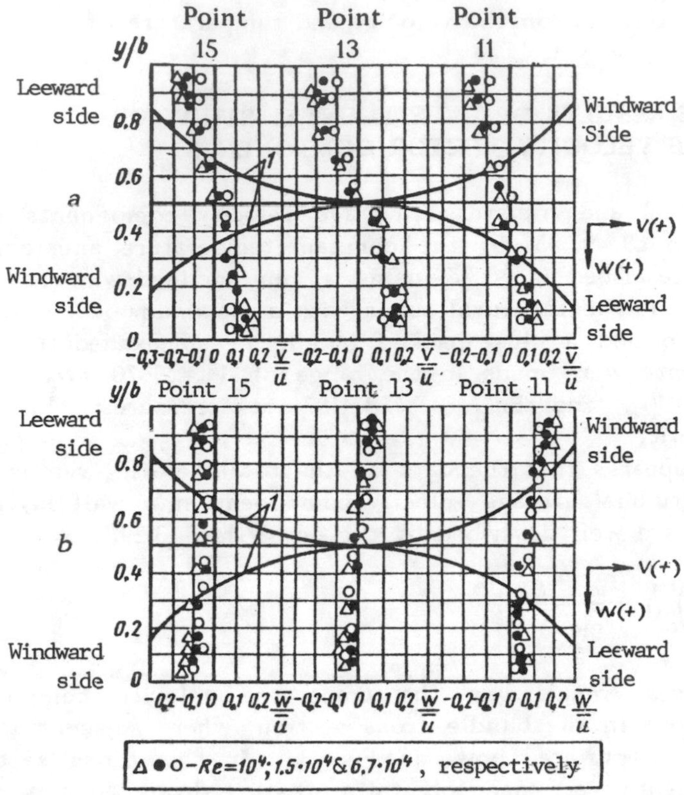

Figure 3.4 Transverse velocity component distributions in bundle cells with $Fr_M = 296$ for $\bar{v}$ (*a*) and $\bar{w}$ (*b*): *1*) conventional boundary of the spiral channel of the tube

If the results on the transverse velocity component distributions (Fig. 3.4) are comapred with data on the fields $\bar{v}$ and $\bar{w}$ in the cross section where the tubes contact over the maximum size of the oval [15], then it is seen that a difference in the velocity distribution $\bar{v}$ is observed to have the same behavior as the velocity distributions $\bar{w}$. Also, in the middle part of the profile [15], the velocity $\bar{w}$ is directed to the wall, while in this part of the profile (Fig. 3.4, point *13*) it is directed from the wall. In the windward part of the tube profile for both cross sections of the bundle, the velocity $\bar{w}$ is directed from the wall and in the leeward part, to the wall with different distributions of $\bar{w}$ over the cell width. This indicates that a helical tube bundle is characterized by a complex spatial vortex motion, and that all the hypotheses considered up to now on the bundle flow pattern are only approximations, whose validity is supported by a coincidence of the predicted and experimental integral characteristics of the flow and heat transfer. However, at present, this approach is probably uniquely feasible because of the difficulties associated with determining a three-dimensional convective heat transfer problem as well as because of the uncertainty of the geometrical flow boundaries in bundles with a great number of arbitrarily oriented helical tubes.

In the peripheral bundle cell, the velocity component fields $\bar{v}$ are determined by helical tube swirling flow and tube shell wall shear. Therefore, in the flow core the velocity direction along the entire tube shell perimeter is the same. At the same time, the velocity $\bar{v}$ changes in magnitude along the tube shell wall. The velocity distribution $\bar{w}$ in the flow core in a peripheral cell is, in magnitude and direction, similar to $\bar{w}$ in the spiral channels of the central cells of the bundle cross section. The dimensionless velocity component $\bar{v}/\bar{u}$ decreases with increasing Re and Fr_M. Thus, its maximum value at Fr_M = 296 decreases from 15–20% at Re = $5.8 \cdot 10^3$ to 7–10% at Re = $6.7 \cdot 10^4$. The maximum value of $\bar{v}/\bar{u}$ decreases from 15–20 to 5–10% when Fr_M increases from 178 to 1187 at Re = $1.5 \cdot 10^4$. The maximum velocity component $(\bar{v}_{max}/\bar{u})_{mean}$, mean over the bundle cross section, as a function of the characteristic similarity numbers, can be described by the following interpolation formula [9]:

$$\left(\frac{\bar{v}_{max}}{\bar{u}_{mean}}\right) = \frac{1.24}{Re^{0.275}}\left(1 + \frac{23.3}{Fr_m} + \frac{31700}{Fr_m^2}\right) \qquad (3.44)$$

The dimensionless velocity component $\bar{w}/\bar{u}$ also decreases with increasing Re and Fr_M, and its maximum value is determined by the relation:

$$\left(\frac{\bar{w}_{max}}{\bar{u}_{mean}}\right) = \frac{0.44}{Re^{0.213}}\left(1 + \frac{95.7}{Fr_m} + \frac{18\,100}{Fr_m^2}\right) \qquad (3.45)$$

Let us consider the velocity component distributions u_τ and u_r as a function of the radial coordinate r from the helical tube coordinate (Fig. 3.5) with respect to typical azimuthal directions around the perimeter of the oval for the tubes of the central and peripheral rows of a bundle.

The tangential velocity component distributions u_τ indicate very complex velocity distributions over the cross section of

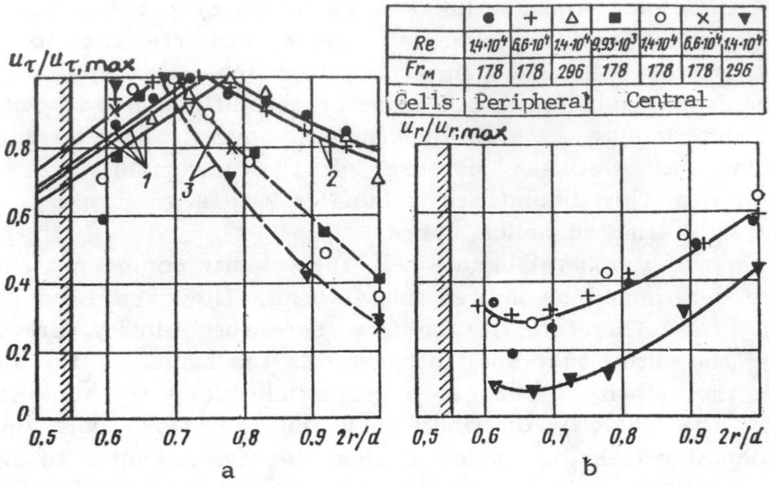

Figure 3.5 Dimensionless profiles of tangential (*a*) and radial (*b*) velocities: *1*) Eq. (3.1); *2*) relation $u_\tau r$ = const; *3*) line for the scatter of the experimental data on the central cells.

the tube bundle [8]. First, the distribution u_τ in the spiral tube channels and in the intertube through channels, into which the entire flow region in a helical tube bundle may be divided, is different. In spiral channels, the direction of the velocity u_τ is specified by the spirally twisted tubes. Therefore, in the external part of the wall layer of all spiral channels, the direction of the velocity u_τ is determined by the tube twisting, and u_τ increases almost linearly over the tube radius (Fig. 3.5). To analyze the data, this allows using, to a first approximation, the hypothesis on flow swirling according to the solid-law, which is satisfied in the case of flow in vortex cords (3.1).

In the flow core of the spiral channels of the central bundle cells, the flow is characterized by various velocity distributions u_τ over the tube radius for different azimuthal directions. This may be explained by the fact that the velocities u_τ appearing in the central cells are initiated by the system of adjacent helical tubes. Thus, for the direction along the tube axis and the straight intertube channel axis, the velocity u_τ (Fig. 3.5) changes from $u_{\tau\max}$ near the external boundary of the wall layer to zero along the straight channel axis. At the conventional boundary of the through channel and the adjacent spiral channels, the velocities u_τ are finite and have an identical direction about the through channel axis, which is responsible for the rotational motion in the through channel. The velocity u_τ in the flow core of a spiral channel for the direction along the tube axis and the through channel axis more strongly decreases over the tube radius rather than following the constant flow circulation law: $u_\tau w$ = const. For the direction along the axis of adjacent tubes, the velocity u_τ in the flow core decreases still more rapidly and is equal to zero at the conventional boundary of the spiral channels of the adjacent tubes. This is bound up with the opposite direction of flow swirling in the adjacent spiral channels (Figs. 3.4 and 3.5).

In the flow core of the spiral channels of a tube bundle's peripheral cells, the velocity u_τ varies similar to the law $u_\tau r$ = const (Fig. 3.5). This may be attributed to the fact that the velocities u_τ appearing in the cells near the tube shell walls are practically initiated by only one helical tube. In all the peripheral bundle cells, the velocities u_τ have the same direction, thereby resulting in a flow circulation in this flow region and relative to the bundle axis. This promotes convective transfer within a bundle diameter.

Thus, the vortex flow in the external part of the wall layer may be considered as the flow in a "vortex cord" of radius $(h/2) + \delta$ (h is the minimum size of the oval), where the velocities u_τ are distributed according to Eq. (3.1). The stream in the flow core may be approximately considered as one in the vicinity of a system of "vortex cords" where all the velocities u_τ are initiated by "vortex cords".

Figure 3.5 shows the radial velocity component as a function of the dimensionless tube radius in the azimuthal direction connecting the axes of the tube and the through channel, where the spiral channel has a maximum size with respect to a radius. At $r = (h/2) + \delta$ the velocity u_r is extremum (as well as the velocity u_τ). The behavior of the velocity u_r is also affected by the specific features of the flow past the contact points of the adjacent tubes.

These studies mainly point out that the theoretical flow models considered in § 3.1 may be used for thermal and hydraulic calculations of helical tube bundles with regard to interchannel heat carrier mixing despite the spatial behavior of the flow in these bundles. This is possible because these specific features of the flow are accounted for in the experimentally determined integral characteristics.

3.4 STATIC PRESSURE DISTRIBUTION IN THE CROSS SECTION OF A HELICAL TUBE BUNDLE

Static pressure distributions over the perimeter of an oval tube (d=36 mm) using 15 pressure gauges were measured on a 19-tube model of a bundle [15]. Pulse tubes, via which a measured signal was supplied to the recording instruments, were located inside a helical tube. Measurements were made with respect to a singular profile point dividing the profile into windward and leeward sides in a bundle cross section where there were no contact points of adjacent helical tubes, i.e., the oval tube profile was streamlined on all sides (Fig. 3.6).

Static pressure distributions over the bundle radius were determined on a 37-tube model of a heat exchanger at d = 46 mm, where the static pressure tubes were inserted through the outlet part of the bundle into the through channels between adjacent helical tubes. The static pressure tubes were fastened to a coordinate mechanism and could be placed at different distances from the heat carrier inlet and at different bundle radii.

The measured velocity distributions in the bundle cross section permitted the assumption that, on the average, the static pressure over a bundle radius and in separate bundle cells changes slightly.

If a helical tube were replaced with a circular cylindrical vortex $(h/2) + \delta$ in radius and the assumption is made that the motion occurs around concentric circles with a constant radius-dependent velocity, then a liquid particle would have only centripetal acceleration equal to:

$$g_c = u_\tau^2/r$$

From Euler's equation, in the projection onto the r-axis:

$$-\rho\ (u_\tau^2/r) = -\partial p/\partial r$$

or

$$p = p_{\max} - \int_r^{d/2} \frac{\rho u_\tau}{r}\, dr \qquad (3.46)$$

it follows that the pressure decreases when approaching the vortex center. Therefore, using the velocity components $\bar{v}$ and

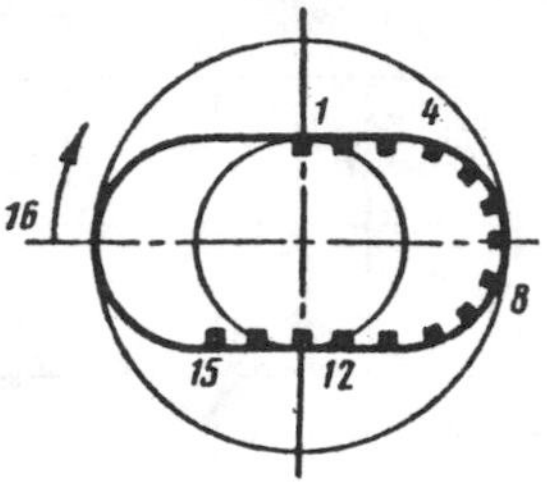

Figure 3.6 Location of static pressure holes around the helical tube perimeter: *1–15*() number of pressure holes; *1–8*) leeward side; *8–15*) windward side; *16*) direction of tube twist

$\bar{w}$ (Fig. 3.4) measured in a cell and the Euler equation of form (3.46), it may be shown that the value of the cell static pressure change with respect to the helical tube channel radius (based on a velocity head $\rho\ \bar{u}^2/2$) is approximately 2%.

The static pressure over the bundle radius was measured in experiments with a jet spreading in a bundle (§ 4.3). This case of flow was considered first because it was characterized by a considerable variation of velocity across the bundle radius, and therefore a static pressure change might be expected.

In this case, the main jet section alone was considered. If the static pressure at a given point is referred to a mean static pressure in this cross section, then it appears that the scatter of the experimental data does not exceed 2% at r/r_{shell} = 0.167, 0.735, and 1.0 at different distances from the bundle inlet (x/l = 0.515, 0.6, 0.74, and 0.9) [15]. It may therefore be concluded that the condition $\partial p/\partial r \approx 0$, or $p = \text{const}(r)$, which was assumed in obtaining the system of equations (3.8)–(3.11), is satisfied, to a first approximation, for helical tube bundles.

Thus, the studies made of the transverse velocity component distributions and of the static pressure enable the use, within the framework of the homogenized flow model, of the equation of motion (3.8) written for a longitudinal velocity component.

The static pressure distribution over the helical tube perimeter (Π_{tube}) was measured on the large-scale model of a heat exchanger. Figure 3.6 presents the location of static pressure holes, and Figure 3.7 illustrates the tube perimeter distribution of the static pressure $\Delta p = p_8 - p_4$) with respect to singular point 8 of the tube profile (Fig. 3.6) dividing the profile into windward and leeward sides. The quantities Δp are referred to a velocity head $\rho u^2_{\text{mean}}/2$. As seen from Fig. 3.7, a

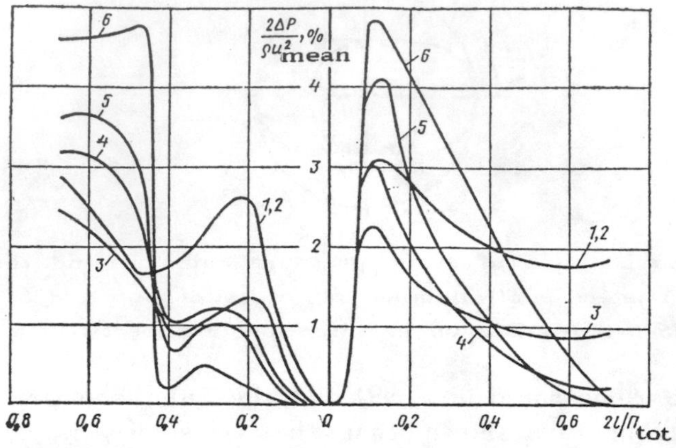

Figure 3.7 Static pressure drop distribution around the tube perimeter based on a velocity head: *1)* Re = 1.1 · 10^5; *2)* Re = 6 · 10^4; *3)* Re = 2 · 10^4; *4)* Re = 7.8 · 10^3; *5)* Re = 6 · 10^3; *6)* Re = 4.5 · 10^3

maximum of the static cell pressure of the bundle cross section, where the tubes do not contact and all cells are connected with the adjacent ones via the slotted channels, is observed at this singular point. A minimum of the static pressure on the windward side of the profile is observed at a distance of $2l/\Pi_{tube} \approx 0.1$, where l is the coordinate over the tube perimeter.

At $2l/\Pi_{tube} > 0.1$ the static pressure increases. Two pressure minima are observed at $2l/\Pi_{tube} \approx 0.2$ and 0.5–0.6 on the leeward side of the profile.

The degree of static pressure nonuniformity over the tube perimeter depends on Re and is very different for the turbulent and transition flow regions of the heat carrier. At Re ≥ $2{\cdot}10^4$ the pressure minima and rates of increasing pressure decrease with increasing distances from the singular point of the profile. With decreasing Re, starting from Re = $7.8{\cdot}10^3$, the pressure minima strongly increase, as does the nonuniformity of the static pressure distribution over the tube perimeter. In the leeward profile region, the minimum near the singular point decreases with decreasing Re. At Re ≥ $6{\cdot}10^4$ the behavior of the pressure distribution and its values remain practically constant (Fig. 3.7). With decreasing Re up to $2{\cdot}10^4$, the nonuniformity of the perimeter distribution of pressure changes slightly, and with a further decrease in Re it sharply increases.

The tube shell perimeter variation of the velocity $\bar{w}$, by its value and direction, exerts an influence on the measured static pressure and, hence, on the accuracy of determining ξ. If the rarefaction regions (Fig. 3.7) are compared with the fields $\bar{w}$ (Fig. 3.4), then it is seen that at point 15 the velocity $\bar{w}$ is directed toward the wall and the static pressure is close to a maximum at point 8. In the profile region, where the velocity $\bar{w}$ is directed from the wall, considerable rarefaction is observed, which is larger, the higher is the value of $\bar{w}$ (points 11 and 13). The largest rarefactions are observed at small Re, when the turbulence intensity is large (Fig. 3.4) and amounts to 5% of the value of the velocity head $\rho u^2_{mean}/2$. A minimum value of $2\Delta p/\rho u^2_{mean}$ is observed at Re = 10^4 - $2{\cdot}10^4$. This quantity slightly increases (Fig. 3.7) with a further increase in Re.

When static pressure gauges are located at individual points over the perimeter of the helical tube and the tube shell, there may appear an additional error in determining p, Δp, and ξ as well as a change in the slope angle of the relation $\xi = \xi$ (Re) (see Chapter 5). It is obvious that the curves $\xi = \xi$ (Re) have an inflection near Re $\approx 8{\cdot}10^3$ - 10^4 as well as the curve $2\Delta p/\rho u^2_{mean} - f(\mathrm{Re})$. The effect of location of the static pressure gauges must be taken into account in correlating the data on ξ obtained by different authors.

Thus, a tube perimeter variation of static pressure and its dependence on Re enable one to account for some specific features of the thermal and hydraulic characteristics. Indeed, a change in $2\Delta p/\rho u^2_{mean} = f \cdot (\Pi_{tube})$ at Re = $7.8{\cdot}10^3$ - $2{\cdot}10^4$ coincides with that in Nu = Nu(Re,Fr_M) and $\xi = \xi$ (Re,Fr_M) for a bundle with the same Fr_M number over the same Reynolds number range (see Chapter 5). This is probably connected with different intensities of exchange of fluid amounts between the wall layer and the flow core in the turbulent and transition flow regions. The Reynolds number, at which there occurs a transition to developed turbulent flow, agrees with the Re at which the inflection of the curves Nu = Nu(Re) and $\xi = \xi$(Re) and the minimum nonuniformity of the static pressure distribution over the helical tube perimeter are observed, thereby initiating the turbulence intensity in a bundle cell.

The experimentally measured static pressure distributions in a helical tube bundle have supported a use of a design homogenized flow model to a first approximation. It also appears that any heat transfer improvement in a helical tube bundle is, to a certain degree, affected by the behavior of the transverse velocity component and the static pressure distributions, which account for the specific features of the thermal and hydraulic characteristics depending on characteristic similarity numbers.

3.5 ENERGY TURBULENCE SPECTRA AND LONGITUDINAL PULSATIONAL VELOCITY COMPONENT DISTRIBUTIONS

A longitudinal pulsational velocity component in a helical tube bundle was studied experimentally by a hot-wire anemometer, and the energy turbulence spectra by a frequency narrow-band analyzer with a working frequency band from 20 to 20000 Hz and frequency resolution of 3 Hz over the range of $Re = 6{\cdot}10^3 - 1.1{\cdot}10^5$ and Fr_M = 178–1187 (s/d = 9.7–25) [8, 9, 15, 16].

These studies enabled one not only to describe the fine flow structure but also to estimate the turbulence intensity and integral turbulence scale in a helical tube bundle. In these studies, the frequency of the signal discretization was determined, to which the hot-wire anemometers were tuned to cover the required frequency range. Agreement between the experimental data on 90 and 28 mm diameter tubes and Laufer's results led to the conclusion that the hot-wire anemometer [9] might be used to investigate the quantity $\bar{u}'^2$ in helical tube bundles. The longitudinal pulsational velocity component was determined using a straight hot wire anemometer, and calculations were made by Eq. (2.21).

Let us estimate the turbulence intensity in helical tube bundles through the averaged velocity components. Assume that in a helical tube bundle, as in the case of circular tube flow, the generation and dissipation of turbulence energy are approximately identical and turbulence is in an energy equilibrium state. However, in a helical tube bundle the turbulence will be generated not only due to a velocity gradient in the wall layer but also due to the total velocity vector nonuniformity in the flow core. Assuming that the turbulence in a tube bundle is isotropic and that the turbulence intensity due to the velocity gradient near the wall is equal to that at a circular tube center

$$v_1/u_{max} \approx 0.044 \tag{3.47}$$

the effect of the velocity nonuniformity in the flow core is allowed for as follows. The value of the pulsational velocity due to the flow core nonuniformity v_2 must be proportional to an excess velocity Δu_{max}. At Re=8·10³, Δu_{max} = 0.4 u_{max} $_{mean}$. In this case,

$$v_2=0.2\Delta u_{max} \tag{3.48}$$

where Δu_{max} is the excess velocity equal to the difference in the minimum and maximum values of the flow core velocity u. A constant multiplier of 0.2 in Eq. (3.48) characterizes the turbulence level in the jet flow, which to a first approximation is observed in the flow core in a helical tube bundle. The total turbulent energy of unit fluid mass will be

$$\frac{3v_\Sigma^2}{2}=\frac{3v_1^2}{2}+\frac{3v_2^2}{2} \tag{3.49}$$

Then, allowing for Eqs. (3.47) and (3.48) and dividing the LHS and RHS of Eq. (3.45) through by u_{max}, from (3.49) we have:

$$v_\Sigma/u_{max}=0.0905 \tag{3.50}$$

Thus, this estimate indicates that the velocity nonuniformity in the flow core is a substantial factor initiating additional flow turbulization in a helical tube bundle.

Experimental results on the longitudinal pulsational velocity components are shown in Fig. 3.8. In the flow core, the dimensionless pulsational velocity distributions along the slots of a helical tube bundle with Fr_M = 178 at Re = 3.6·10⁴ may be described by a sine wave whose period is equal to the maximum size of a tube profile (Fig. 3.9):

$$\frac{\sqrt{\overline{u'^2}}}{u_{max}}=0.075+0.025\sin 2\pi\,\frac{x-x_0}{d} \tag{3.51}$$

In Eq. (3.51) the origin of $x_0 \approx \delta$ is on the tube shell wall and the quantity u_{max} is the local average velocity in the flow core. The maxima of curve (3.51) fall beyond the contact points of the tubes nearest to the outlet cross section where the measurements were made, and the minima in the through (free of blades) part of the channel. Contact of helical oval tubes is illustrated by the example of a heat exchanger (see § 1.1).

The measured values $\sqrt{u'^2}/u_{max}$ across the slots near the maximum of the curve (Fig. 3.10) have shown that the behavior

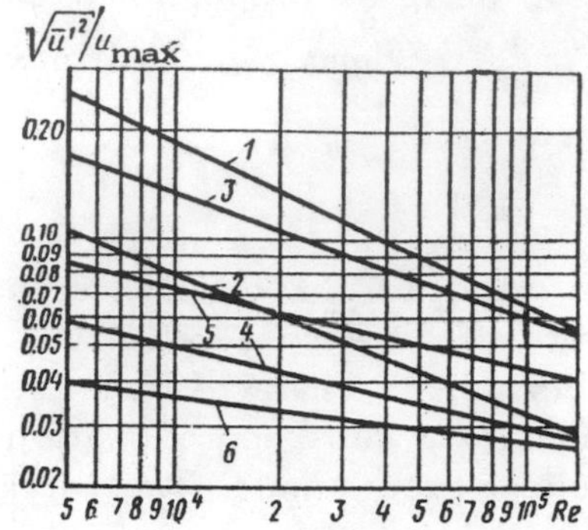

Figure 3.8 Relative longitudinal pulsational velocity component versus Re and Fr_M at typical flow points: *1, 2*) experimental data for Fr_M = 178 at the points with maximum and minimum values of $\sqrt{u'^2}/u_{max}$, respectively, described by Eq. (3.52); *3, 4*) the same for Fr_M = 296; *5, 6*) the same for Fr_M = 1187

of $\sqrt{u'^2}/u_{max}$ across the slots is similar to that of a relative longitudinal pulsational velocity component in a circular tube [54], and $\sqrt{u'^2}/u_{max}$ in a tube bundle in the flow core and near the wall is higher than the one in the circular tube. Thus, the value of the turbulence intensity in a tube bundle at a distance $2y/d_{eq}$ = 0.01 from the wall is 30%, and in a circular tube 15% [54]. In the region of the through channel, $\sqrt{u'^2}/u_{max}$ is constant and equal to 7%.

These investigations have shown that at Re < 10^5 in the flow core of a helical tube bundle the turbulence intensity is, on the average, higher than in a circular tube, which is

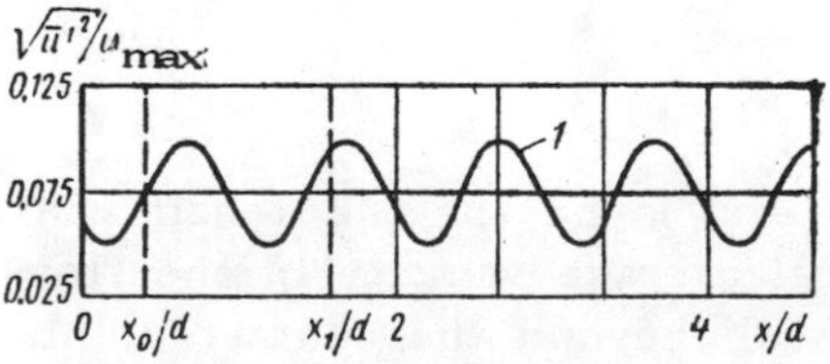

Figure 3.9 Longitudinal pulsational velocity distribution in the flow core along the tube rows at Fr_M = 178 and Re = $3.6 \cdot 10^4$: *1*) Eq. (3.51).

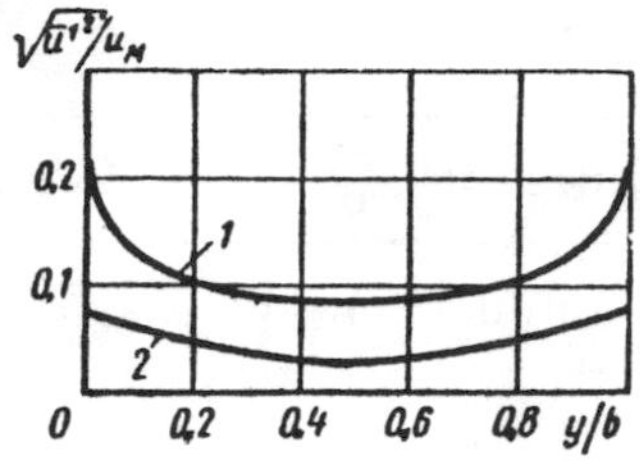

Figure 3.10 Longitudinal pulsational velocity distribution across the tube rows (from wall to wall of the wide side of the oval profile of adjacent tubes) at Fr_M = 178 and Re = $3.6 \cdot 10^4$: *1)* experimental data; *2)* circular tube data

consistent with the above estimate made using the turbulent energy balance equation. As mentioned above, the increase in $\sqrt{u'^2}/u_{max}$ in a tube bundle, as compared to a circular tube, is attributed to flow turbulization due to the velocity field nonuniformity in the flow core. The sinusoidal behavior of changing $\sqrt{u'^2}/u_{max}$ is due to the existence of contact points of the tubes. The velocity u_{max} behind the nearest contact point of the tubes, close to the cross section where measurements have been made, is smaller than in the region of the through channel. If the measured quantity $\sqrt{u'^2}$ is referred to the velocity u_{max} mean averaged over the cross section, then the quantity $\sqrt{u'^2}/u_{max.mean}$ along the slot may vary over smaller limits. Analysis of the effect of Re and Fr_M on the quantity $\sqrt{u'^2}/u_{max}$ (Fig. 3.8) has indicated that in the range of Re = $6 \cdot 10 - 1.1 \cdot 10^5$ the experimental data can be described by the formula:

$$\sqrt{\overline{u'^2}} \,/\, u_{max} = A/Re^m \tag{3.52}$$

The values of A_0 and m in Eq. (3.52) are given in Table 3.1.

If the stream in the flow core is characterized by the quantity $\sqrt{u'^2}/u_{max}$ averaged over the cross section of a bundle, then the following relation [9] may be assumed:

$$\frac{\sqrt{\overline{u'^2}}}{u_{max}} = 7.2\,Re^{-\left(0.155+40.57\,Fr_M^{-1}+1700\,Fr_M^{-2}\right)}\left[1+\frac{Fr_M-178}{7.5\,(19.5-0.135\,Fr_M)}\right] \tag{3.53}$$

Table 3.1

Fr_M number	Values of A_o max	min	mean	Values of m max	min	mean
178	12.85	2.96	7.5	0.46	0.393	0.437
296	3.26	0.403	1.62	0.347	0.227	0.311
1187	0.61	0.104	0.313	0.230	0.115	0.189

which is valid over the range of Re = $6\cdot10^3$–$1.\cdot10^5$ and Fr_M = 178–1187. Equation (3.53) at Re = $8\cdot10^3$ may be represented as

$$\frac{\sqrt{\overline{u'^2}}}{u_{max}} = 0.044\,(1 + 368\,Fr_M^{-1}) \tag{3.54}$$

which satisfactorily represents the experimental data on the effective turbulence intensity at Fr_M = 314 (§ 4.2). This intensity was determined by the method of heat diffusion from a point source, assuming that the flow core turbulence was isotropic and uniform.

Thus, the turbulence intensity in a helical tube bundle increases with decreasing Re and Fr_M. These results are supported by data [14], in which the largest heat transfer enhancement, as compared to a circular tube, is observed at Re < 10^4. The data obtained differ greatly from those repeated in [29, 35], where it is shown that over the range of Re = $8\cdot10^3$–10^8 the turbulence intensity in a flat channel practically does not depend on Re. This difference is attributed to the specific features of the flow structure in a spirally twisted tube bundle.

It should be noted that Eqs. (3.53) and (3.52) may be used mainly to estimate the turbulence level in a helical tube bundle. They indicate that, as compared to a smooth tube bundle, substantial flow turbulization is one of the main reasons for heat transfer improvement and flow mixing in helical tube bundles.

The quantity $\sqrt{\overline{u'^2}}$ may be expressed in the form of a one-dimensional energy turbulence spectrum:

$$\overline{u'^2} = \int_0^\infty E(f)\,df \tag{3.55}$$

or, if we pass from a frequency f to a wave number k:

$$k=2\pi f/u \tag{3.56}$$

then in the form of a spectrum we have:

$$\overline{u'^2}=\int_0^\infty E(k)\,dk \tag{3.57}$$

The value of $E(k)$ may be determined from measurements using a normalized spectrum $E(k)/\overline{u'^2}$ because the quantity $\overline{u'^2}$ is known from hot-wire anemometer measurements and data processing by (2.21).

Study of the energy spectrum (3.57) in a helical tube bundle allows simultaneously determining the longitudinal integral turbulence scale, necessary to estimate the turbulent diffusion coefficient, and the frequency distribution of the kinetic vorticity-dependent turbulence energy. If the turbulence considered incorporates only large vortices, then the function $E(f)$ will be substantial in the low-frequency region, and if it contains small vortices, then this function will be strong in the high-frequency region. Thus, an analysis of the spectra, $E(f)$, enables determining to what type of turbulence the considered flow is referred and what region in a bundle is occupied by energy-containing vortices which possess maximum kinetic energy in a developed turbulent flow. Moreover, since the energy dissipation affected by viscosity increases with decreasing vortex size and attains a maximum value for small vortices, the energy spectrum shift to the high-frequency region suggests a reason for the increasing hydraulic resistance coefficient, which is observed in a helical tube bundle compared to circular tubes. The need to study energy spectra is also explained by the methodological reasons associated with tuning hot-wire anemometers when used to examine the quantity $\overline{u'^2}$.

Energy spectra were investigated at two points of the flow core corresponding to the maximum and minimum values of $\sqrt{\overline{u'^2}}/u_{max}$ (Fig. 3.8) [16].

The results on the energy spectra $\overline{u'^2}$ behind the nearest point of contacting adjacent tubes in the coordinates $E(f) = \psi(f)$ are shown in Fig. 3.11. Here, for comparison, the energy spectra for a circular tube are plotted. The results of discrete measurements are replaced by solid lines in Fig. 3.11. From this figure it is seen that the flow in helical tube bundles, as

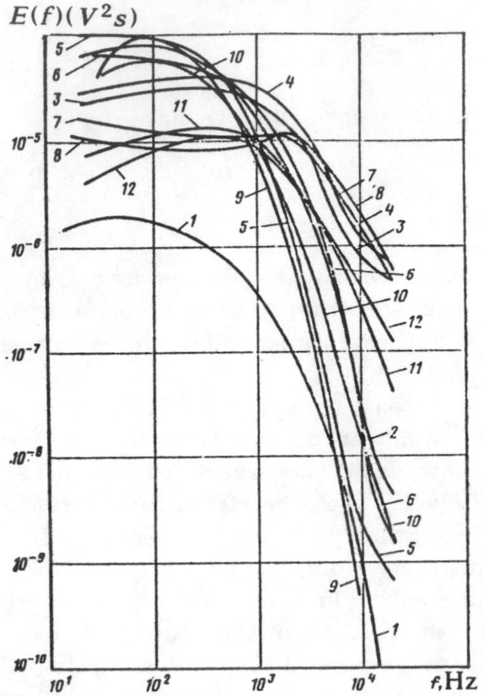

Figure 3.11 Spectral distribution of longitudinal pulsational velocity components for the flow core point corresponding to the maximum value of $\bar{u}'^2/u_{max}$: *1)* experimental data for the tube axis d_{tube} = 90 ;and 28 mm at Re = 5.6 · 10^3–1.31 · 10^5; *2)* experimental data for a bundle with Fr_M = 178 at Re = 1.2 · 10^4; *3)* experimental data for a bundle with Fr_M = 178 at Re = 3.4 · 10^4; *4)* experimental data for a bundle with Fr_M = 178 at Re = 7.5 · 10^4; *5, 6, 7, 8)* the same for a bundle with Fr_M = 296 at Re = 6.7 · 10^3, 1.4 · 10^4, 4 · 10^4, and (6.7–11) · 10^4, respectively; *9, 10, 11, 12)* the same for a bundle with Fr_M = 1187 at Re = 8.08 · 10^4, 1.41 · 10^4, 4.05 · 10^4, and 6.1 · 10^4, respectively.

compared to that in a circular tube, is characterized by a shift of energy turbulence spectra to the high-frequency region. Moreover, the spectral density over the frequencies of relative turbulence energy transferred by a component u' depends on Fr_M and Re. A contribution of the high-frequency spectrum components to the quantity $\overline{u'^2}$ increases with decreasing Fr_M and increasing Re. Such a change of energy turbulence spectra in helical tube bundles accounts for the behavior of the hydraulic resistance coefficient as a function of Fr_M and Re [34, 14]. Indeed, in a helical tube bundle the hydraulic resistance coefficient increases with Fr_M, and an increment of this coefficient increases with increasing Re, starting from some critical value ($Re_{cr} \approx 10^4$), as compared to a circular tube.

In processing the experimental data in the form of the relation $E(k) = \varphi(k)$, the data for different Re at a certain Fr_M number are practically described by one relation (Fig. 3.12).

From Fig. 3.12 it is seen that in processing the spectra with respect to wave numbers, the dependence of the spectra $E(k)$ on Fr_M is preserved, as in the case of data processing in the coordinates $E(f) = \psi(f)$.

Energy spectra obviously characterize the vortex turbulence structure. At $Re < Re_{cr}$, energy in a helical tube bundle is mainly transferred by low-frequency large-scale vortices. A maximum frequency of energy-containing vortices is approximately 500 Hz (Fig. 3.11). The same result on the frequency level was observed in [61] for turbulent flow in a circular tube bundle (Fig. 3.12). When the Reynolds number is increased, the situation abruptly changes, especially for bundles with large Fr_M numbers. At $Re > Re_{cr}$, the interval of the frequencies of energy-containing vortices is extended approximately up to 2-3 kHz, and the similarity of the frequency distribution of relative turbulence energy is observed at 20-1000 Hz (Fig. 3.11). The results for energy spectra are qualitatively the same for the points behind a contact point of adjacent tubes and in a through channel.

Study of the energy turbulence spectra in a helical tube bundle enables one to determine the frequency range within which the measuring equipment must be tuned depending on Re and Fr_M. It appears that the equipment must be tuned at frequencies f = 0-50 kHz to study $\overline{u'^2}$ in a helical tube bundle, while in investigating $\overline{u'^2}$ in a circular tube the equipment must be tuned at f = 0-20 kHz.

It has proved possible to estimate the longitudinal integral turbulence scale through the measured energy spectra (Fig. 3.12) using the Taylor hypothesis on frozen turbulence expressed as:

$$L_1 = \frac{\pi}{2\overline{u'^2}} \lim_{k \to 0} E(k) \qquad (3.58)$$

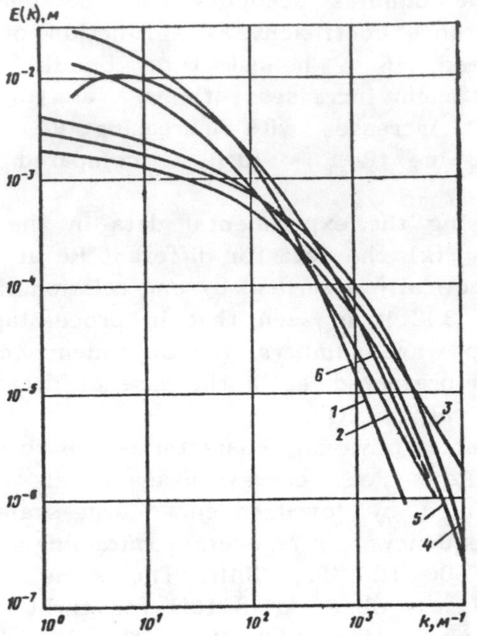

Figure 3.12 Energy spectra of u' for the flow core point corresponding to the maximum value of $\sqrt{\overline{u'^2}}/u_{max}$: *1)* Laufer's data; *2)* experimental data for tubes with d_{tube} = 90 and 28 mm at Re = $6.3 \cdot 10^4$–$1.13 \cdot 10^5$; *3)* experimental data for Fr_M = 178 at Re = $1.2 \cdot 10^4$–$7.5 \cdot 10^4$; *4)* the same for Fr_M = 296 at Re = $6.7 \cdot 10^3$–$6.7 \cdot 10^4$; *5)* the same for Fr_M = 1187 at Re = $8.08 \cdot 10^3$–$6.1 \cdot 10^4$; *6)* data of [61]

This quantity determines a mean statistical size of vortices. It is found that a turbulence scale L_1 depends on the Fr_M number and on the position of a measuring point in the flow core. At points with a minimum value of $\overline{u'^2}$ (through channel), L_1 is smaller than L_1 at points with a maximum value of $\overline{u'^2}$. $L_{1\ mean}$ is introduced as an arithmetical mean of $L_{1\ max}$ and $L_{1\ min}$ for a given Fr_M number to characterize the flow on the average. Then, for $L_{1\ mean}$ the following relation may be assumed:

$$L_{1\ mean}/d_{eq} = 2.41\ Fr_M^{-0.407} \tag{3.59}$$

The considered energy turbulence spectra also allows an estimation of the validity of the law:

$$E(k) \sim k^{-5/3} \tag{3.60}$$

in helical tube bundles for the region characterized by the vortices that appear in the equilibrium flow at large Reynolds numbers. Since the spectrum $E(k)$ was studied for a longitudinal pulsational velocity component, the wave number domains where (3.60) was valid for different Fr_M were determined. It appears that if law (3.60) exists with changing k over the range covering more than half of the order, then this law is satisfied only for a bundle with Fr_M = 178 behind a contact point of adjacent tubes. For all other considered bundles (Fr_M = 296 and 1187), as well as in the region of the through channels of a bundle with Fr_M = 178, law (3.60) is not satisfied, which is consistent with Laufer's data for tube flow. This indicates that at small Fr_M, starting from some value of this number, the wake effect behind the contact points of adjacent tubes causes qualitative changes in the flow structure in a helical tube bundle.

From these studies it follows that the turbulence in a helical tube bundle is generated not only by the tube wall impact but also by tangential discontinuities when jets are spreading with different velocities and in different directions. The jet effect is especially pronounced at small Fr_M numbers and is, within this range, responsible for a sharp increase in transfer coefficients.

CHAPTER
FOUR

CROSSFLOW MIXING IN CLOSE-PACKED HELICAL TUBE BUNDLES IN LONGITUDINAL FLOW

4.1 THEORETICAL MODELS FOR FLOW MIXING. METHODS OF SOLVING A SYSTEM OF EQUATIONS FOR A HOMOGENIZED FLOW MODEL

Different theoretical mixing models may be used to calculate the velocity and temperature fields of the heat carrier in tube bundles with a uniform field of heat supply or energy release. In [4], consideration was made of the homogenous flow method adopted to determine two-dimensional temperature fields and transverse and longitudinal velocity vector components in a heat exchanger with a lateral supply of an incompressible liquid. In [49], the temperature fields in tube bundles were calculated in a two-dimensional statement by the model for an anisotropic porous body with volumetric friction and heat release sources. A solution of the hydrodynamic equations is sought in the form of the stream-function and the vorticity function. However, these methods cannot be employed to calculate three-dimensional temperature fields in heat exchangers with spirally twisted oval-shaped tube bundles using air or other substances as heat carriers, whose density varies with changing temperature and pressure.

The method of calculating elementary cells with regard to the effects of mass, momentum, and energy transfer between them is most widely used to determine three-dimensional temperature fields. The intensity exchange between heat carrier flows in adjacent cells of a bundle is characterized, in this case, by a flow mixing coefficient, which is a transferred flow fraction per unit bundle length with respect to the total channel flow:

$$\mu = G_{ij}/G_i \qquad (4.1)$$

However, in this case, when the number of tubes is large, considerable computer time is needed to implement the computational program. Therefore, in the present investigation, the homogenization method of a real helical tube bundle in a heat exchanger is adopted to calculate three-dimensional temperature and velocity fields. The flow of a homogenized medium in a heat exchanger is described by the continuity equation, and thc homogenization effect is taken into account by including the multiplier $(1-m)/m$. The porosity m may be determined by allowing for a boundary layer displacement thickness δ^*. In this case, consideration is given to free flow involving the slip of a homogenized medium with distributed sources of volumetric heat release and hydraulic resistance within the geometrical limits of the flow with regard to δ^*. In [48], a study was made of the flow of some two-phase medium where a solid phase is movable, and in [44], the flow with a fixed solid phase was studied within the geometrical limits of the flow with no regard to the thickness δ^*. A difference in the interpretation of the homogenization model [48, 44] leads to systems of equations written either for the flow core, or for mean mass parameters of the flow. However, according to [25], in helical tube bundles, in which the velocity and temperature field profiles are more full, as compared to an equivalent circular channel, the difference between the flow core and mean mass parameters is not large and, in a number of cases, does not exert a noticeable effect on the calculation results.

To calculate three-dimensional velocity and temperature fields of the heat carrier, whose density is a function of temperature and pressure, the system of equations for flow and heat transfer that governs the homogenized flow in a helical tube bundle may be written in the form:

$$\overline{\rho}\overline{u}\frac{\partial \overline{u}}{\partial x}=-\frac{dp}{dx}+\frac{1}{r}\frac{\partial}{\partial r}\left(r\nu_{ef}\frac{\partial \overline{u}}{\partial r}\right)+\frac{1}{r^2}\frac{\partial}{\partial \varphi}\left(\nu_{ef}\frac{\partial u}{\partial \varphi}\right)-\xi\frac{\overline{\rho}\overline{u}^2}{2d_{eq}} \quad (4.2)$$

$$\overline{\rho}\overline{u}c_p\frac{d\overline{T}}{\partial x}=q_v\frac{1-m}{m}+\frac{1}{r}\frac{\partial}{\partial r}\left(r\lambda_{ef}\frac{\partial \overline{T}}{\partial r}\right)+\frac{1}{r^2}\frac{\partial}{\partial \varphi}\left(\lambda_{ef}\frac{\partial \overline{T}}{\partial \varphi}\right) \quad (4.3)$$

$$G=m\int_0^{2\pi}\int_0^{r_{shell}}\rho\overline{u}r\,dr\,d\varphi \quad (4.4)$$

$$p=\overline{\rho}R\overline{T} \quad (4.5)$$

In writing Eqs. (4.2)-(4.5), the assumption was made that the convective terms in the equations of motion due to transverse and azimuthal velocity components may be ignored. Equations (4.2)-(4.5) are supplemented with boundary conditions:

$$\left.\begin{aligned} &\bar{u}(r, \varphi, 0)=\bar{u}_{in}(r, \varphi) \\ &\bar{T}(r, \varphi, 0)=\bar{T}_{in}(r, \varphi) \\ &p(0)=p_{in} \end{aligned}\right\} \tag{4.6}$$

$$\left.\frac{d\bar{u}}{dr}\right|_{r=r_{shell}}=0, \quad \left.\frac{\partial \bar{T}}{\partial r}\right|_{r=r_{shell}}=0 \tag{4.7}$$

$$\begin{aligned} &\bar{u}(r, \varphi, x)=\bar{u}(r, \varphi+2\pi, x) \\ &\bar{T}(r, \varphi, x)=\bar{T}(r, \varphi+2\pi, x) \end{aligned} \tag{4.8}$$

The system of equations (3.8)-(3.11) with boundary conditions (3.12)-(3.14) has been solved to determine the axisymmetric velocity and temperature fields of the heat carrier.

The effective turbulent viscosity coeffciient ν_{ef} and the turbulent thermal conductivity λ_{ef} in Eqs. (4.2) and (4.3) allow for all the transfer mechanisms in a helical tube bundle: turbulent diffusion, convective transfer due to the eddy motion in bundle cells, and ordered transfer in the spiral channels of the tubes. The peripheral cells of a bundle are also characterized by the eddy motion relative to the tube bundle axis, which must result in additional enhancement of a transfer process at azimuthal nonuniformity of heat supply. The quantities ν_{ef} and λ_{ef} may be expressed in terms of the effective turbulent diffusion coefficient, D_t, assuming that the turbulent Lewis and Prandtl numbers are equal to unit ($Le_T = \rho c_p D_t/\lambda_{ef} = 1$, $Pr_T = c_p \nu_{ef}/\lambda_{ef} = 1$):

$$\lambda_{ef}=D_t\bar{\rho}c_p \tag{4.9}$$

$$\nu_{ef}=\bar{\rho}D_t \tag{4.10}$$

It is more convenient to operate with a dimensionless effective diffusion coefficient:

$$k=D_t/\bar{u}d_{eq} \tag{4.11}$$

which depends on the characteristic similarity numbers Re, $Fr_M = s^2/dd_{eq}$, etc. This coefficient should be determined from experiment both for the azimuthal nonuniformity and for the axisymmetric nonuniformity of the heat supply.

The system of equations (4.2)-(4.5) and the appropriate boundary conditions is a system of nonlinear parabolic-type equations. The solution method and algorithm of such a system, which can be used to calculate the three-dimensional temperature fields in multitube bundles, are detailed in [44]. When the coefficient k in a helical tube bundle was determined by this method, theoretical calculations were made at some prescribed values of this coefficient. Thus, the design temperature fields of the heat carrier formed a grid within which were the measured temperatures of the heat carrier [25]. The experimentally measured value of Q and the volume of the region of a helical tube bundle, where a group of heated tubes was mounted, were used to prescribe a volumetric heat release source q_V. In the calculations, use was also made of the experimentally measured velocity and temperature fields of the heat carrier at the inlet of the design section of the bundle that modelled azimuthal nonuniformity of the heat supply in a heat exchanger with a lateral heat carrier supply.

The governing system of equations (4.2)-(4.5) was solved numerically, the numerical analogs of the equations being written implicitly using the matrix factorization method and using iteration cycles with respect to the nonlinearities. In utilizing this method, the greatest difficulty was associated both with writing the finite-difference analogs of the governing equations at a singular point on the helical tube bundle axis ($r=0$) and with including the conditions for azimuthal periodicity of the unknown functions into one of the coefficient matrices.

Gershgorin's method [11], which enables expressing the value of an unknown function on the bundle axis through a set of azimuthal values on the first design radius, was adopted to write the numerical analogs of the governing equations on the bundle axis. The periodicity condition was introduced as follows. To solve the finite-difference analogs of the energy and motion equations, these were reduced to the form [25]:

$$A_i\psi_{i+1} - B_i\psi_i + C_i\psi_{i-1} = -F_i \tag{4.12}$$

where ψ_i is the set of azimuthal values of a desired function on the i-th design radius. In this case, the matrices A_i, B_i, and C_i are composed of the coefficients of the equations. To implement the periodicity conditions, the tridiagonal matrix B_i was supplemented with angular elements which automatically allowed for this condition on all design radii of the domain of determining the unknown functions.

It should also be noted that as a preliminary step, the equation of motion was split into two equations using Simuni substitution [46] to eliminate the pressure gradient:

$$\overline{u}_{i,j}=\overline{w}_{i,j}+z_{i,j}(dp/dx) \tag{4.13}$$

For each layer along the bundle length, a pressure gradient was determined from the integral relation for the heat carrier flowrate with respect to the helical tube bundle cross section:

$$\frac{dp}{dx}=\frac{G-m\int_0^{2\pi}\int_0^{r_{shell}} u\rho r dr d\varphi}{m\int_0^{2\pi}\int_0^{r_{shell}} z\rho r dr d\varphi} \tag{4.14}$$

The solution algorithm of the governing system of equations (4.2)-(4.5) with their boundary conditions was implemented in the form of a FORTRAN computer program [44]. The program allows calculating the values of the heat carrier temperature and velocity at 1500 nodes of a spatial grid for 12-13 minutes when thermophysical properties of the heat carrier depend on the flow parameters. This illustrates the rather high response speed of this program.

The nonlinear system of parabolic-type equations for the axisymmetric problem (3.8)-(3.11) was solved by the grid method using an explicit scheme [43]. After the dimensionless variables

$$\hat{T}=\frac{\overline{T}}{\overline{T}_{in}},\ \hat{u}=\frac{\overline{u}}{\overline{u}_{in}},\ \hat{p}=\frac{p}{p_{in}},\ \hat{r}=\frac{r}{r_{shell}},\ \hat{x}=\frac{x}{L} \tag{4.15}$$

were included, the system of equations (3.8)-(3.11) was reduced to a dimensionless form and was written in finite differences [12]. In this case, the constants c_1 through c_6 entering the finite-difference analogs of the governing differential equations had the form:

$$C_1 = k\,\frac{L}{r_{shell}}\,\frac{d_{eq}}{r_{shell}} \tag{4.16}$$

$$C_2 = \frac{Lq_V\,(1-m)}{\overline{\rho}_{in}\,\overline{u}_{in}\,c_p\,\overline{T}_{in}\,m} \tag{4.17}$$

where

$q_V = Q/\pi r^2 Lm$

$$C_3 = C_1 \, Pr_T \tag{4.18}$$

$$C_4 = \xi \frac{L}{2d_{eq}} \tag{4.19}$$

$$C_5 = \frac{p_{in}}{\bar{\rho}_{in} \, \bar{u}_{in}^2} \tag{4.20}$$

$$C_6 = 2\pi m \bar{\rho}_{in} \, \bar{u}_{in} r^2_{shell} \tag{4.21}$$

Then, the number of nodes N on the r-axis was introduced, a step h_r was determined, and an assumption made that $h_x = 1/5 \; h_r^2$. The number of steps with respect to the radius and height were chosen for reasons of accuracy and minimum computer time. For the considered problem, these conditions were satisfied by a choice of 1355 steps over the height and 82 steps over the radius. Then, the equations of motion and energy may be written as:

$$\hat{u}(\hat{x}+h_x, \hat{r}) = \hat{u}(\hat{x}, \hat{r}) + z_u(\hat{x}, \hat{r}) h_x + C_5 \frac{\hat{T}(\hat{x}, \hat{r})}{\hat{u}(\hat{x}, \hat{r})} \frac{\hat{p}(\hat{x}+h_x) - \hat{p}(\hat{x})}{\hat{p}(\hat{x})} \tag{4.22}$$

$$\hat{T}(\hat{x}+h_x, \hat{r}) = \hat{T}(\hat{x}, \hat{r}) + z_T(\hat{x}, \hat{r}) h_x \tag{4.23}$$

where

$$z_u(\hat{x}, \hat{r}) = C_3 \frac{\hat{T}(\hat{x}, \hat{r})}{\hat{u}(\hat{x}, \hat{r})\hat{r}} D\left(\hat{u}, \frac{\hat{r}}{\hat{T}} \hat{u}\right) - C_4 \hat{u}(\hat{x}, \hat{r}) \tag{4.24}$$

$$z_T(\hat{x}, \hat{r}) = C_1 \frac{\hat{T}(\hat{x}, \hat{r})}{\hat{u}(\hat{x}, \hat{r}) \hat{r}} D\left(\hat{u} \frac{\hat{r}}{\hat{T}}, \hat{T}\right) + C_2 \frac{\hat{T}}{\hat{p}\hat{u}} \tag{4.25}$$

The functions $D\left(\hat{u} \; \hat{r}/\hat{T}, \; \hat{u}\right)$ and $D\left(\hat{u} \; \hat{r}/\hat{T}, \; \hat{T}\right)$ in (4.24) and (4.25) may be expressed in a general form as:

$$D(f, \varphi) = \frac{\partial}{\partial r}\left(f \frac{\partial \varphi}{\partial r}\right)$$

or in finite difference form

$$D(f,\varphi)=\frac{1}{2h_r^2}[(f_{l+1}+f_l)(\varphi_{l+1}-\varphi_l)-(f_l+f_{l-1})(\varphi_l-\varphi_{l-1})] \tag{4.26}$$

The pressures at point $\hat{x}+b_x$ were found from the continuity equation:

$$G_0=\hat{p}(\hat{x}+h_x)\int_0^1 \frac{\hat{u}(\hat{x}+h_x,\hat{r})\hat{r}d\hat{r}}{\hat{T}(\hat{x}+h_x,\hat{r})}=\hat{p}(\hat{x})\int_0^1 \frac{\hat{u}(\hat{x},\hat{r})\hat{r}d\hat{r}}{\hat{T}(\hat{x},\hat{r})} \tag{4.27}$$

The system of equations (4.22) and (4.27) can then be simplified, assuming that $\hat{T}(\hat{x}+h_x,\hat{r})$ does not depend on $\hat{p}(\hat{x}+h_x)$. $\hat{u}(\hat{x}+h_x,\hat{r})$ linearly depends on $\hat{p}(\hat{x}+h_x)$. Introducing

$$G_1=\int_0^1 \frac{u'(\hat{x}+h_x,\hat{r})}{T'(\hat{x}+h_x,\hat{r})}\hat{r}d\hat{r} \tag{4.28}$$

$$G_2=\int_0^1 \frac{\hat{T}(\hat{x},\hat{r})}{\hat{T}(\hat{x}+h_x,\hat{r})\hat{u}(\hat{x},\hat{r})}\hat{r}d\hat{r} \tag{4.29}$$

yields

$$\delta p=\frac{G_1-\dfrac{G_0}{p(x)}}{G_2-\dfrac{G_1}{C_5}} \tag{4.30}$$

The velocity data for the values of the argument $\hat{x}+h_x$ is

$$\hat{u}(\hat{x}+h_x,\hat{r})=u'(\hat{x}+h_x,\hat{r})-\frac{\hat{T}(\hat{x},\hat{r})}{\hat{u}(\hat{x},\hat{r})}\delta p \tag{4.31}$$

and the pressure at a point $\hat{x}+h_x$ is

$$\delta p=C_5\frac{p(\hat{x}+h_x)+p(\hat{x})}{p(\hat{x})} \tag{4.32}$$

The boundary conditions for u' are of the form:

$$u'\,\hat{x}+h_x,r_{shell})=u'(\hat{x}+h_x,r_{shell}-h_r) \tag{4.33}$$

The above method was employed to construct an algorithm and the program was implemented on a computer [12]. The geometrical characteristics of a tube bundle $(F,\ m,\ d,\ r_H,$

r_{shell}, L), pressure, temperature and velocity of the heat carrier at the bundle inlet, the quantity q_v, as well as the physical properties of the heat carrier were prescribed to make the calculations. The temperature and velocity fields in a helical tube bundle were calculated using this program. This calculation accuracy was checked by comparing a mean mass heat carrier temperature at the bundle outlet with one obtained for a predicted temperature field in the outlet cross section of the bundle. The accuracy associated with determining these temperatures was < 1%.

4.2 RESULTS OF THE EXPERIMENTAL STUDY OF TRANSPORT FLOW PROPERTIES EMPLOYING THE METHOD OF DIFFUSION FROM A POINT HEAT SOURCE

Studies were made on models consisting of 37 helical tubes 46 mm diameter, 1000 mm long, and with d_{eq} = 24.6 and 11.5 mm using the heat diffusion method in the experimental set-up described in § 2.4. An indicator gas, namely, heated air, was continuously supplied via a circular tube fitted on the central helical tube of the bundle into the cocurrent longitudinal cold airflow around the bundle. The chosen scale of diffusion source provided indicator gas supply. The amount of the indicator gas was 2% of the main air flowrate, which allowed its effect on turbulence to be neglected. The indicator gas flowrate was chosen in a manner to ensure equality of the velocities of the indicator gas and main flows. Temperature and velocity fields in the outlet cross section of a bundle were measured by a chromel-alumel thermocouple with a bead 0.5 mm in diameter and by a total pressure 1.2 mm diameter pressure gauge 0.1 mm in wall thickness which were mounted on a coordinate screw mechanism so that the diffusion source might travel relative to this cross section. The use of these methods was substantiated experimentally.

Experiments were conducted over a range of Re = $4.3 \cdot 10^3 - 8.2 \cdot 10^3$, at Fr_M = 314 and 1530, temperatures $T_{max} \leq 340$ K, temperature drops $T_{max} - T_0$ = 10–450, and at a pressure close to atmospheric.

The Reynolds number was determined assuming that d_{eq} was taken as the characteristic dimension; the mean mass temperature of the heat carrier was taken as a characteristic temperature; and a mean mass velocity in the tube bundle was taken as a characteristic velocity.

Experimental studies of flow transport properties were made in a tube bundle arranged both in ordered and in random fashion. It appears that the tube arrangement in a bundle practically does not influence the temperature distribution over a bundle radius.

The tubes of the bundle were made of aluminium-base alloy. A series of experiments were run with helium used as an indicator gas to estimate the effect of thermal conductivity of the tubes on the measured temperature fields and the determined transfer characteristics. Coincidence of the transfer characteristics measured by the methods of heat and mass diffusion as well as design estimates have shown that such an effect is not substantial.

Typical measured dimensionless excess temperature fields in the cross section of a helical tube bundle located in succession from a diffusion source at Fr_M = 314 are shown in Fig. 4.1, where these are compared with the estimated distribution for a point source and for a source of finite size.

The following design scheme (Fig. 4.2) was adopted to allow for the finite sizes of the diffusion source. Diffusion from a ring shaped source r_0 in radius was considered. In this case, a local temperature increase at point N in the flow at distance y from point M of the ring will be proportional to $\exp(-y^2/2\bar{y}^2)$. Then, the local temperature increase at point N due to heat diffusion from all the points of the ring-shaped source will be proportional to:

$$T-T_0 \sim 2\int_0^{\pi} \exp\left[-\frac{r_0^2 \sin^2 \alpha + (r - r_0 \cos \alpha)^2}{2\bar{y}^2}\right] d\alpha \tag{4.34}$$

Designating $\cos\alpha = x$ and $rr_0/\bar{y}^2 = a$, upon some transformations we have:

$$T-T_0 \approx 2\exp\left(-\frac{r_0^2 + r^2}{2\bar{y}^2}\right)\int_{-1}^{+1} \frac{\exp(ax)\,dx}{\sqrt{1-x^2}} \tag{4.35}$$

The integral in expression (4.35) may not be expressed in terms of elementary functions. Therefore, let us use a series expansion integration of the integrand $\exp(ax)$:

$$\exp(ax) = 1 + \frac{ax}{1!} + \frac{a^2x^2}{2!} + \frac{a^3x^3}{3!} + \dots \tag{4.36}$$

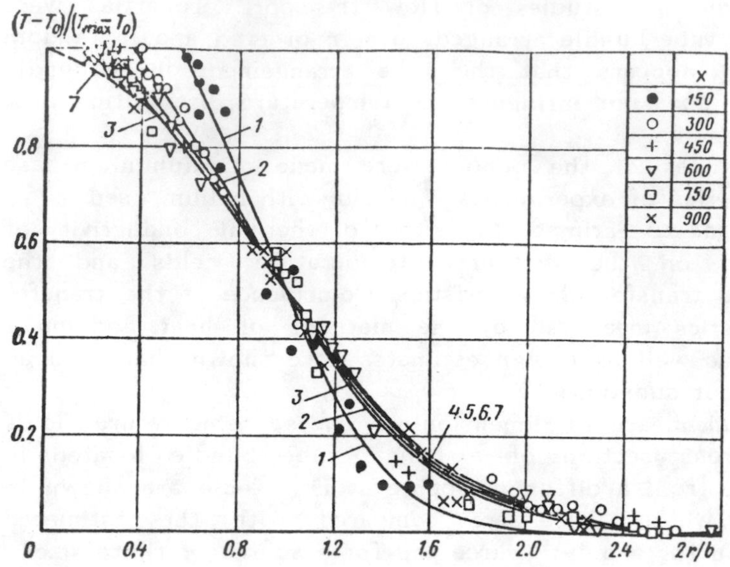

Figure 4.1 Comparison of the experimentally measured and calculated distributions of dimensionless excess temperatures for a helical tube bundle with Fr_M = 314: *1–6*) calculation by Eq. (4.39) for x=150, 300, 450, 600, 750, and 900 mm, respectively; *7*) calculation by Eq. (4.40)

The expansion series converges at $|x| < \infty$. Upon restriction to six terms of the series expansion (4.36), from Eq. (4.35) we approximately arrive at:

$$T-T_0 \sim 2\pi \exp\left(-\frac{r_0^2+r^2}{2\bar{y}^2}\right)\left(1+\frac{r^2r_0^2}{4(\bar{y}^2)^2}+\frac{r^4r_0^4}{64(\bar{y}^2)^4}+\frac{r^6r_0^6}{2304(\bar{y}^2)^6}\right) \tag{4.37}$$

Dividing expression (4.37) through by a quantity $T_{max} - T_0$

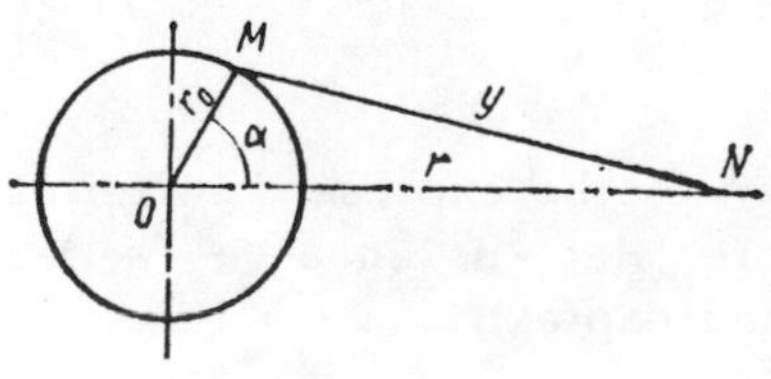

Figure 4.2 Design scheme of a ring-shaped source

$$T_{max} - T_0 \sim 2\pi \exp\left(-\frac{2r_0^2}{2\bar{y}^2}\right)\left(1 + \frac{r_0^4}{4(\bar{y}^2)^2} + \frac{r_0^8}{64(\bar{y}^2)^4} + \frac{r_0^{12}}{2304(\bar{y}^2)^6}\right) \tag{4.38}$$

and using the dimensionless quantities $\bar{r}_0 = 2r_0/b$, $\bar{r} = 2r/b$, and $\bar{\bar{y}}^2 = \bar{y}^2/b^2$ yields:

$$\frac{T - T_0}{T_{max} - T_0} = \exp\left(\frac{\bar{r}_0^2 - \bar{r}^2}{8\bar{\bar{y}}^2}\right)$$
$$\times \frac{\bar{r}^6\bar{r}_0^6 + 576(\bar{\bar{y}}^2)^2\bar{r}^4\bar{r}_0^4 - 147{,}5\cdot 10^3(\bar{\bar{y}}^2)^4\bar{r}^2\bar{r}_0^2 + 9{,}45\cdot 10^6(\bar{\bar{y}}^2)^6}{\bar{r}_0^{12} + 576(\bar{\bar{y}}^2)^2\bar{r}_0^8 + 147{,}5\cdot 10^3(\bar{\bar{y}}^2)^1\bar{r}_0^4 + 9{,}45\cdot 10^6(\bar{\bar{y}}^2)^6} \tag{4.39}$$

At $\bar{r}_0 = 0$, Eq. (4.39) is identical to a Gaussian distribution for a point source on the bundle axis:

$$\frac{T - T_0}{T_{max} - T_0} = \exp\left(-\frac{r^2}{2\bar{y}^2}\right) \tag{4.40}$$

which is a particular solution of the energy equation:

$$\rho u c_p \frac{\partial T}{\partial x} = \frac{1}{r}\frac{\partial}{\partial r}\left(\rho r c_p D_t \frac{\partial T}{\partial r}\right) \tag{4.41}$$

Equation (4.41) describes the heat diffusion from a point source in a uniform flow for the homogenized flow model. In this case, the problems of turbulent and molecular diffusion in isotropic bodies are identical.

The results calculated by formula (4.39) at $x \geq 600$ mm are in good agreement with the experimental data and calculation by formula (4.40) (Fig. 4.1). A small difference is observed only near the distribution peak. Based on this, to determine the effective diffusion coefficient by Eq. (2.25), the quantity $\bar{y}^2$, distribution (4.40), is

$$\bar{y}^2 = 0{,}179b^2 \tag{4.42}$$

At small distances from a diffusion source, a good deal of discrepancy is found between the results calculated by Eqs. (4.40) and (4.39) at $x \leq 150$ mm. Therefore, use of the considered method of determining the effective turbulence intensity allows one to make only an approximate estimate of the quantity ε.

Figure 4.3 gives the values of $\bar{y}^2$ calculated by Eq. (4.42) for the experimentally measured temperature distributions at different distances from the diffusion source. Using the limiting solutions of Taylor's equation, (2.24) and (2.25), the relation in Fig. 4.3 allows determination of the quantities ε and D_t/ud_{eq} which are equal to 0.116 and 0.09, respectively, at Fr_M = 314, and equal to 0.094 and 0.073, respectively, at Fr_M = 1530 (with Re ≈ 8000).

Utilizing the measured values of ε and assuming that at $Fr_M \to \infty$ $\varepsilon \approx 0.044$, an equation is obtained for calculation of the effective turbulence intensity:

$$\varepsilon=0.044\,(1+8.1\,Fr_M^{-0.278}) \tag{4.43}$$

The formula for calculating the effective diffusion coefficient at a turbulent Prandtl number Pr_T = 1 may be obtained from the relationship

$$D_t=\chi l_1 v_1 \tag{4.44}$$

where

$$v_1=\sqrt{\overline{u'^2}}=\sqrt{\overline{v'^2}}=\sqrt{\overline{w'^2}} \tag{4.45}$$

if it is assumed that in the flow core the pulsational velocities along the coordinate axes are equal. In a dimensionless form, expression (4.44) is of the form:

$$\frac{D_t}{uR_{shell}}=\chi\,\frac{l_1}{R_{shell}}\,\varepsilon \tag{4.46}$$

where r_{shell} is the tube bundle radius and r_{shell}/d_{eq} = 6.55.

The dimensionless mixing path for the flow region in a tube bundle was assumed to be equal to $l_1/r_{shell} \approx 0.1$, which is typical of the external part of a wall layer and of jet flows [1]. The proportionality factor χ was determined from a comparison of the predicted and the experimental data on the diffusion coefficient. For the analyzed flow core χ = 1.235. When Eq. (4.43) was determined for the quantity ε in Eq. (4.46), an assumption was made that at $Fr_M \to \infty$ the turbulence intensity in a straight oval tube bundle was equal to that at the external boundary of the wall layer for straight tube flow, $\varepsilon \approx 0.044$. Then, substituting into expression (4.46), instead of ε, its value from (4.43), we obtain the relation

$$k_{asy}=\frac{D_t}{ud_{eq}}=0.0356\,(1+8.1\,Fr_M^{-0.278}) \tag{4.47}$$

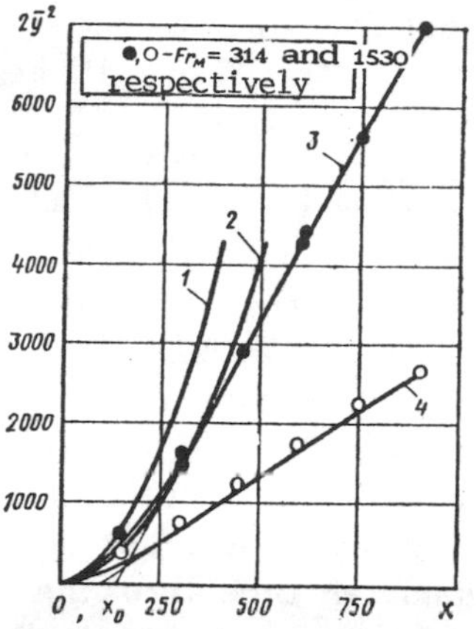

Figure 4.3 Experimental doubled mean-static square of the displacement $2y^2$ versus distance up to a diffusion source x: *1, 2*) Eq. (2.24); *3–4*) Eq. (2.25)

which is in fair agreement with the experimental data. In this series of experiments, the Reynolds number had no observed effect on the coefficient k_{asy}.

4.3 JET SPREADING IN A HELICAL TUBE BUNDLE

Knowledge of the specific features of jet spreading in helical tube bundles is required for estimating the entrance length needed to analyze the experimental heat carrier mixing data obtained by the diffusion method from linear heat sources. In this case, the converted longitudinal tube bundle coordinate $2_{ax}/d$, where the jet structure coefficient a must be found from experiment, is included to allow for the effect of tube bundle length on the diffusion coefficient D_t. In Tolmin's theory, this coefficient is the only empirical constant and allows generalization of the experimental data on the maximum velocity attenuation along a free axisymmetric jet outflowing from an orifice with different-varying velocity profiles as well as with turbulizing grids and swirlers located at the jet mouth [1].

Strictly speaking, the basic assumptions and the laws of free jet flow in a helical tube bundle are not satisfied, and experimental justification is necessary to use them, even approximately. In a helical tube bundle the jet spreads and interacts with the tube and cassette walls. Within the framework of the model for a homogenized medium at $Pr_T = 1$, the equation of motion for this flow will be of the form:

$$\rho u \frac{\partial u}{\partial x} + \rho v \frac{\partial u}{\partial r} = -\frac{dp}{dx} + \frac{1}{r}\frac{\partial}{\partial r}\left(\rho r D_t \frac{\partial u}{\partial r}\right) - \xi \frac{\rho u^2}{2d_{eq}} \tag{4.48}$$

From Eq. (4.48) it is seen that the diffusion term $1/r\ \partial/\partial r$ $(\rho r D_t\ \partial u/\partial r)$ and the term $\xi\ \rho u^2/2d_{eq}$, responsible for the wall friction impact, describe the temperature levelling effect on the jet in a helical tube bundle.

The jet spreading in a helical tube bundle was studied by the method of a moving jet orifice. The air jet was injected via a movable 46 mm diameter tube into the central helical tube of a bundle. Helical oval 1000 mm long tubes with a maximum d = 46 mm profile were oriented in a bundle so that the symmetry axes of the tube profiles were parallel to each other in the outlet tube bundle cross section where the velocity fields were measured. The jet in a tube bundle was spreading in the heat exchanger intertube space bounded by the cassette walls. The velocity fields were measured by a 1.2 mm diameter 0.1 mm-thick wall Pitot tube mounted on a coordinate screw mechanism.

The experimental set-up, measuring system, and experimental procedures were checked by investigating a circular jet using this set-up. For this purpose the helical tubes were taken out from the cassette to allow the circular jet to spread without hindrance inside the cassette. The jet outflowing from the tube had the following parameters: a cut velocity of 35 m/s, Re $\approx 10^5$, and a velocity nonuniformity in the jet mouth $\approx$ 1.22. As is known, in a range of Re = $0.2 \cdot 10^5 - 40 \cdot 10^6$, the variation of the maximum axial velocity along the jet does not depend on Re [1]; therefore, the study was made at a fixed Reynolds number. The investigation of circular jet spreading showed that the experimental distributions of the dimensionless excess velocity in the cross sections of the main region of a jet were consistent with the Gauss curve, and the varying maximum axial jet velocity at a = 0.0757 was described by the hyperbolic relation of Tolmin's theory using the method of a moving jet orifice.

Jet spreading in helical oval tube bundles was investigated at Fr_M = 80, 317, and 1560, bundle porosity $m \approx$ 0.4–0.5 with respect to the heat carrier, and at Re $\approx 10^5 - 1.5 \cdot 10^5$. The Reynolds number was determined through the tube outlet parameters.

Static pressure distribution at a fixed position of the jet orifice was analyzed to support the assumption that $p \approx$ const (x,r) was satisfied in the entire bundle volume and in the outlet cross section where the static pressure was equal to the ambient pressure. It turned out that only near the jet mouth was the deviation of the static pressure from ambient $\pm 3\%$, and this was attributed to variation of the flow direction; in the entire bundle volume, where the jet is spreading, the condition $dp/dx \approx 0$ typical of free jets is essentially satisfied.

Figures 4.4 and 4.5 show the data on jet spreading in helical tube bundles. The experimentally measured velocity distributions in successive cross sections of the main jet region are well represented by the following interpolation relation

$$\frac{\Delta u}{\Delta u_{max}} = \exp\left[-2.01\left(\frac{r}{r_{mean}}\right)^2\right] + 0.0288\exp\left[6.75\frac{r}{r_{mean}} - 4.21\left(\frac{r}{r_{mean}}\right)^2\right] \quad (4.49)$$

where $\Delta u_{max} = u_{max} - u_0$ is the maximum excess velocity in the jet cross section and $\Delta u = u - u_0$ is the excess velocity over a given bundle radius r.

Equation (4.49) points both to the similarity of the velocity profiles in the main jet region and to the axisymmetry of the problem. Comparison of velocity profile (4.49) with Schlichting's profile:

$$\frac{\Delta u}{\Delta u_{max}} = \left[1 - \left(0.44\frac{r}{r_{mean}}\right)^{1.5}\right]^2 \quad (4.50)$$

and with the Gauss curve:

$$\frac{\Delta u}{\Delta u_{max}} = \exp\left[-2.79\frac{r^2}{4r_{mean}^2}\right] \quad (4.51)$$

(Fig. 4.4) shows that near the peak of the curve, Eqs. (4.50) and (4.51) give underestimated values of the dimensionless excess velocity, while in the boundary zone they give overestimated ones, i.e., velocity profile (4.49) is more full as compared to profiles (4.50) and (4.51). The fact that profile (4.49) is uniform may be attributed to the imapct of friction forces exerting a levelling effect on the velocity profile. However, since velocity profile (4.49) is, by its nature, a jet type, it may be that the friction force impact on this profile in a helical tube bundle is not pronounced. This fact is also supported by agreement between the experimental data on the

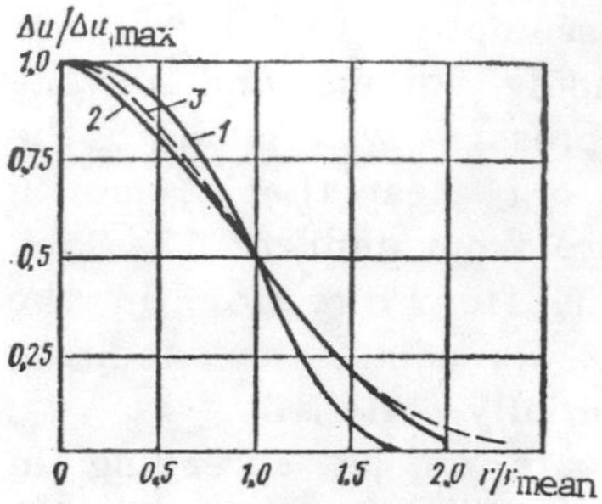

Figure 4.4 Dimensionless excess velocity profile in the cross section of the main region of a jet spreading in a helical tube bundle and its comparison with Schlichting's profile and Gauss' curve: *1*) empirical relation (4.49); *2*) Eq. (4.50); *3*) Eq. (4.51)

varying maximum excess velocity along a jet spreading in a helical tube bundle in its main region (Fig. 4.5) and the theoretical curve for an axisymmetric free jet:

$$\frac{\Delta u_{\max x}}{\Delta u_{\max 0}} = \frac{0.96}{\frac{2ax}{d} + 0.29} \tag{4.52}$$

where $\Delta u_{\max 0} = u_{\max 0} - u_0$ is the maximum excess velocity in the jet mouth (x=0).

As follows from Fig. 4.5, in a helical tube bundle the maximum velocity continuously decreases along the jet, starting from the coordinate $x/d \approx 1$, and then a transition to the

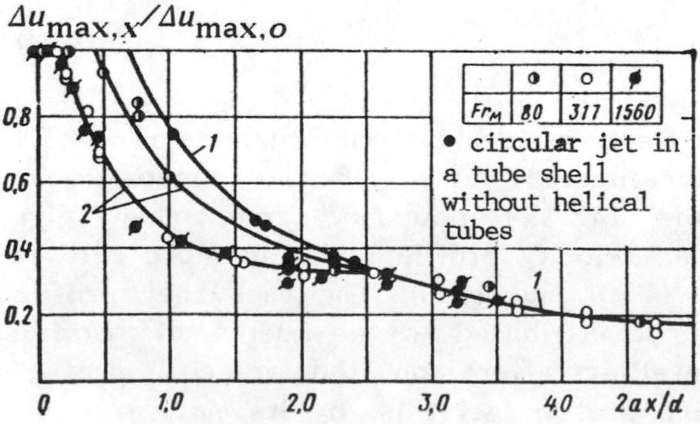

Figure 4.5 Variation of maximum excess velocity along a jet spreading in a helical tube bundle: *1*) Eq. (4.52); *2*) Eq. (4.54)

hyperbolic law for maximum velocity attenuation along the jet, (4.52), occurs when the value of the longitudinal coordinate is determined by the expression:

$$\left(\frac{2ax}{d}\right)_{f'}=2.56 \tag{4.53}$$

In a range of $x/d \approx 1.0$–1.28, when the jet is spreading in a tube bundle, a transition zone is observed and has no analog in the case of a free spreading jet. The variation of the maximum velocity in the transition jet region is governed by the power dependence:

$$\frac{\Delta u_{\max x}}{\Delta u_{\max 0}} = k \frac{0.96}{b\left(\frac{2ax}{d}\right)^{n} + 0.29} \tag{4.54}$$

At $2ax/d$ = 1–2.56, b = 1.87, and n = 0.408, the quantity k = 1 for Fr_M = 317–1560 and $k = 1.41\,(2ax/d)^{-0.365}$ for Fr_M = 80. At $2ax/d$ = 0.2–1.0, b = 1.9, and n = 0.696, the quantity k = 1 for Fr_M = 317–1560 and k = 1.141 for Fr_M = 80.

The transition to the main jet region corresponds to the moment at which the jet penetrates behind the second row of tubes. In this case, formation of the jet is caused by no less than 19 tubes. It may be assumed that in this case, the similarity structure of a jet flow develops.

The jet structure coefficient, a, characterizing the jet attenuation rate, was determined experimentally through a measured maximum velocity $\Delta u_{\max x}$ in the main jet region using Eq. (4.52). The coefficient a determined in this manner has the following values: at Fr_M = 80, a = 0.256; at Fr_M = 317, a = 0.113; at Fr_M = 1560, a = 0.082 and may be described by

$$a=0.0745+11.37\,Fr_M^{-1}+246\,Pr_M^{-2} \tag{4.55}$$

The coefficient a for a helical tube bundle can also be used to generalize the experimental data on varying counterflow velocity u_0 in a circulation bundle flow zone and on the geometrical jet characteristics, r_{mean}. The variation of velocity u_0 (its presence is caused by the jet spreading in the bounded space) along the jet is described, over a range of Fr_M = 80–1560, by the following relation:

$$\frac{u_0}{u_{max}} = -0.1165\left(\frac{2ax}{d}\right)^{0.775} \qquad (4.56)$$

The geometrical jet characteristic, r_{mean}, in a helical tube bundle varies linearly along the jet both for the main jet region

$$2r_{mean}/d = 1.75 + 0.568\,(2ax/d) \qquad (4.57)$$

and for the transition region

$$2r_{max}/d = 0.85 + 0.916\,(2ax/d) \qquad (4.58)$$

The validity of applying the jet laws was also supported by estimates of the mean excess velocity in the jet, which proved to be $(\Delta u_{mean}/\Delta u_{max})_x = 0.411$ (for a submerged axisymmetric jet $u_{mean}/u_{max} = 0.258$ [1], as well as by an estimate of the air flowrate and momentum, which appeared to be almost constant along the jet.

Thus, the study of the jet spreading process in a helical tube bundle has shown that the laws typical of a free jet are generally satisfied, and a definite value of the converted longitudinal coordinate (4.53), at which a stabilized velocity profile develops, may be used to estimate the entrance tube length needed to examine the data on heat carrier mixing.

4.4 RESULTS OF THE EXPERIMENTAL STUDY OF MIXING EMPLOYING THE METHOD OF DIFFUSION FROM A SYSTEM OF LINEAR HEAT SOURCES

In practice, the nonuniformity of the heat carrier temperature field in the intertube space of a helical tube heat exchanger is formed mainly not by point but by linear heat sources of finite sizes, for example, by heat supply via the tube walls which is nonuniform over the bundle cross section. It is, therefore, of interest to study an effective diffusion coefficient in a helical tube bundle using the method of diffusion from a system of linear finite-size heat sources. This method allows the homogenized flow model to be used more successfully than the generally accepted method of heating an identical central tube. In this case, a group of tubes may be segregated in a bundle. These may be heated by electric current. A group of heated tubes is electrically insulated from a bundle of steel tubes. When these tubes are heated, a nonuniformity in the heat carrier temperature field in the bundle will develop and will be partially equalized along the bundle due to cross mixing. The effective diffusion coefficient characterizing heat transfer in a helical tube bundle can be determined from a comparison of the experimentally measured temperature fields of the heat carrier in the outlet cross section and the predicted ones, using methods of mathematical statistics.

In determining an effective diffusion coefficient k, two methods [12] were employed to compare the predicted and experimentally measured temperature fields. The first method implied that for each design curve $\bar{T} = T(\bar{r})$, at a given value of the coefficient k, the square root was evaluated from a sum of the squared deviations of each experimental point from this curve, and the relation

$$\sqrt{\sum_{i=1}^{n}(\delta\bar{T}_i)^2} = f(k) \tag{4.59}$$

was plotted, where n is the number of experimental points.

A minimum of function (4.59) agrees with a maximum reliable value of the dimensionless coefficient k, at which the best agreement between experiment and theory is achieved. The second method consisted of making a statistical analysis to determine confidence ranges of k using experimental sampling. For this, the methods of mathematical statistics were employed. First, the possibility of applying the statistical hypotheses on all of the calculated dispersions belonging to one general dispersion and on a normal distribution of a random quantity $\bar{T}$ was ascertained. Then, a statistical analysis was made of k. A certain value of k was assigned to each experimental point on the plot $\bar{T} = T(\bar{r})$ in accordance with the theoretical curves $\bar{T} = T(k, r)$ drawn on this plot. The intervals of the quantity k were determined with the accepted confidence probability of 0.95. The grid of the design curves $\bar{T} = T(\bar{r})$ on the experimental plot had a number of intervals with respect to k equal to m and an interval length with respect to the quantity k equal to h. Calculation was made in the following manner. A determination was made on the number of points within the interval from k_i to $k_i + h$ (which were assigned to the middle of the interval), on the mean value of the quantity k in a given sample,

$$\bar{k}=\frac{1}{n}\sum_{i=1}^{m} n_i \bar{k}_i \qquad (4.60)$$

and on the corrected dispersion of the quantity k:

$$\sigma^2=\frac{1}{n-1}\sum_{i=1}^{m} n_i(\bar{k}_i-\bar{k})^2-\frac{h^2}{12} \qquad (4.61)$$

Then, considering the hypothesis on the normal distribution of quantity k, the confidence intervals of quantity k will be determined by the formula:

$$k=\bar{k}\pm 1.96\sigma \qquad (4.62)$$

where $\sigma = \sqrt{\sigma^2}$ is the rms deviation of k.

This method was employed to determine the effective turbulent diffusion coefficients for axisymmetric and asymmetric nonuniformities of a heat release field in a bundle cross section.

The Axisymmetric Case The study of cross mixing of a heat carrier with axisymmetric nonuniformity of a heat release field in a bundle of spirally twisted tubes was made on the experimental set-up described in § 2.4. Several tubes were heated in bundles composed of 37 helical tubes 750 mm long, and a group of 37 central tubes was heated in bundles composed of 127 helical tubes 500 and 1500 mm long to develop a uniform field of heat supply for the heat carrier. The tubes had a maximum profile size d = 12.3 mm and the wall thickness was equal to 0.2 mm. The tubes were given a coat of electrically insulating varnish. Air served as the heat carrier. The heat carrier temperature fields were measured at the outlet bundle cross section by means of thermocouples fixed in a coordinate screw mechanism. The airflow entered a helical tube bundle axisymmetrically. Experiments were made both with heat carrier flow from a large volume into a tube bundle, when the turbulence intensity at the bundle inlet was $\varepsilon \approx 1\%$, and with heat carrier flow through a system of three turbulizing grids into a tube bundle, when the turbulence intensity was $\varepsilon \approx 6\%$, to study the effect of inlet turbulence on the coefficient k. The turbulence intensity at the bundle inlet was measured by a constant-temperature hot-wire method [9]. The coordinate screw mechanism was also equipped with a total Pitot tube to measure velocity heads. To facilitate measurements of the temperature and velocity fields, the tubes in the bundle were arranged to form free spaces between their rows in the form of slots in the displacement plane of the measuring probes. The measuring system and the method of mathematical data processing resulted in an ultimate error in determining the coefficient $k = \pm 25$–50%.

Experiments were made at Fr_M = 55–1080, Re = $3.4 \cdot 10^3$–$3.8 \cdot 10^4$, porosity m = 0.477–0.545, and inlet flow turbulence ε = 1–6% [12, 18].

Typical distributions of the experimentally measured heat carrier temperature fields for Fr_M = 232 and Re = $(1.2\text{–}1.7) \cdot 10^4$ are shown in Fig. 4.6. The experimental flow velocity distribution in the outlet bundle cross section is also presented in Fig. 4.6 for a bundle at Fr_M = 232. It is seen that the temperature field nonuniformity found by a nonuniform heat release field over the bundle radius also initiates a longitudinal flow velocity. This points to the fact that in equation of motion (3.8), the diffusional term $1/r\ \partial/\partial r\ (\rho r D_t\ \partial u/\partial r)$ should also allow for the levelling of velocity nonuniformities. In Fig. 4.6, the experimental temperature and velocity distributions are compared with the theoretical distributions at different values of the coefficient k, when the system of equations (3.8)–(3.11) with boundary conditions (3.12)–(3.14) is solved explicitly by the network method. The behavior of the predicted temperature and velocity is similar to that of the experimental distributions, but the experimental and predicted fields coincide at different values of the coefficient k as a function of Fr_M. The smaller the Fr_M number, the larger is the coefficient k at which these fields coincide.

Agreement between the calculated temperature and velocity fields and the experimental ones may form an experimental basis for using the homogenized flow model and the elaborated method of calculating a system of differential equations describing the homogenized flow.

Representation of the temperature fields in a dimensionless form $T = T(r/r_{shell})$ leads to a considerable spreading of the design curves for different values of k at a relatively small scatter of the experimental points. Therefore, experimental data processing in the form $\bar{T} = T(r/r_{shell})$ allow the unknown coefficient k to be determined within a sufficient accuracy. However, such processing requires maintenance of the heat carrier temperature at the bundle inlet, its flowrate, and

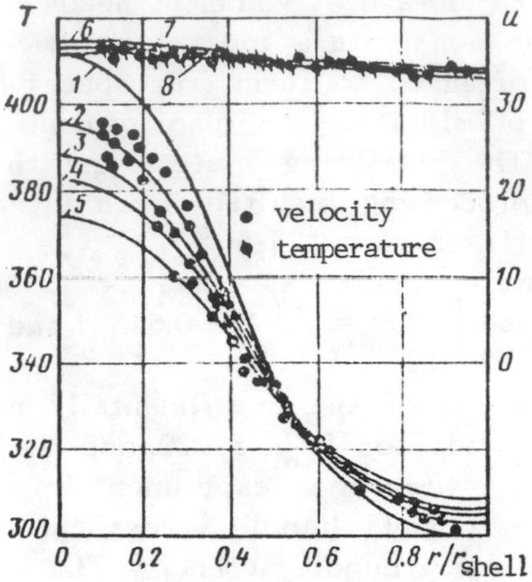

Figure 4.6 Comparison of the experimentally measured temperature and velocity fields with the theoretical calculations for a bundle with Fr_M = 232: *1–5*) temperature calculation at k = 0.03, 0.04, 0.045, 0.05, 0.06; *6–8*) velocity calculation at k = 0.03, 0.045, 0.06

heat power supplied to the heated helical tubes at a prescribed level when statistical values are set up in experiments.

Use of these methods to compare the experimental temperature distribution of the heat carrier and calculations in the form of (4.59) and (4.60) made it possible to determine the effective diffusion coefficients k for axisymmetric nonuniformity of heat release over the entire experimental range of parameters. These data will be analyzed and generalized in the form of the critical relations described in § 4.5.

The Asymmetric Case In the case of azimuthal nonuniformity of heat supply to a helical tube bundle, the effective diffusion coefficient k was determined by the method of electrical heating of azimuthal tubes [25]. Experiments were made with a bundle of 37 oval tubes at Fr_M = 64–1050, Re = 11840–16600, heat power Q = 2.84–9.9 kW, and T_{inlet} = 275–306 K on the experimental set-up used in the axisymmetric case.

These studies emphasize that the coefficient k is constant in the cross section of a helical tube bundle. For this, consideration was given to different shapes of the asymmetric heating zone of the helical tubes: a peripheral zone, shaped as

2-3 rows of tubes located near one side of the hexahedral tube shell, and a zone shaped as a tape of two rows of tubes which do not contact the cassette wall.

Experimentally measured temperature fields of the heat carrier are shown in Figs. 4.7-4.9. These figures show the cross-sectional schematics of the close-packed bundles of the investigated shapes and the relative-length zones where the heated helical tubes were located. They also show the calculated temperature fields of the heat carrier, which are obtained by solving the system of equations (4.2)-(4.5) with boundary conditions (4.6)-(4.8) [25].

Large discrepancies have been found between experiment and theory in all the cases considered only near the helical tubes contacting the tube shell, i.e., in the peripheral region of a bundle. This can probably be attributed to the fact that the elevated porosity of the bundle and the available eddy flow relative to the bundle center cause the local transfer coefficient to substantially exceed the coefficient k for the central regions of the bundle.

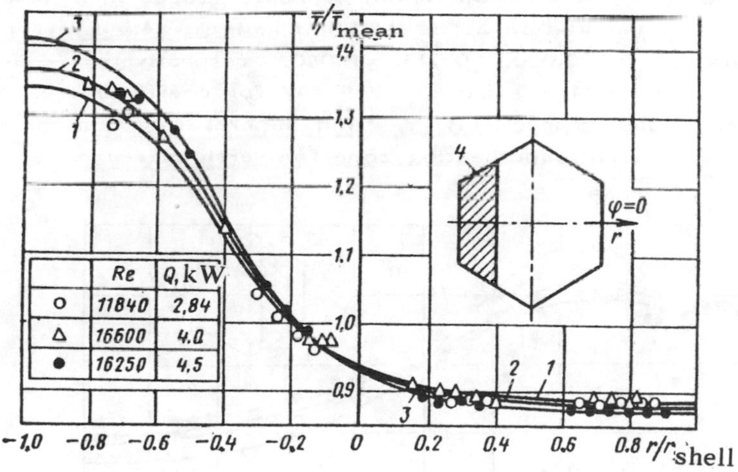

Figure 4.7 Fluid temperature fields for a helical tube bundle with Fr_M = 1050 and for the azimuthal direction φ = 0 at different Reynolds numbers and heat release amounts: *1-3*) calculation at k = 0.03 for Re = 11840 and Q = 2.84 kW, Re = 16,600 and Q = 4 kW, and Re = 16,250 and Q = 4.5 kW, respectively; *4*) region of location of heated tubes in heat exchanger

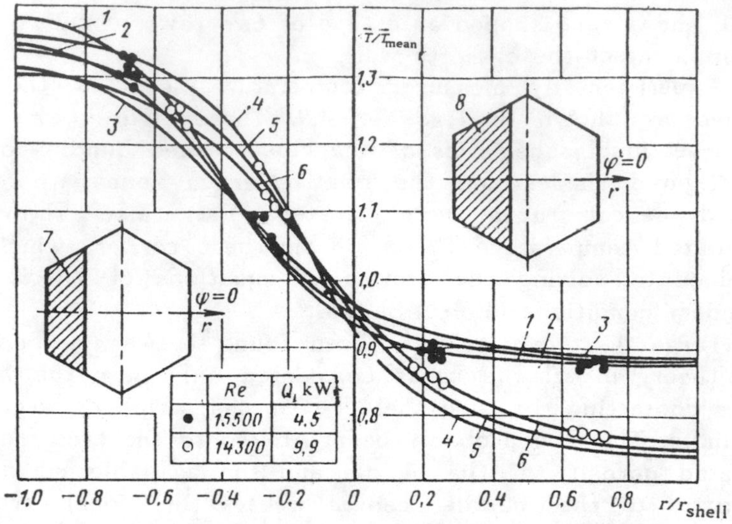

Figure 4.8 Fluid temperature fields for a helical tube bundle with Fr_M = 232 and for the azimuthal direction φ = 0 at different sizes of the location region of heated tubes in a heat exchanger: *1–3*) calculation at Re = 15,500 and Q = 4.5 kW for coefficients k = 0.030, 0.045, 0.060, respectively; *4–6*) calculation at Re = 14,300 and Q = 9.9 kW for coefficients k = 0.03, 0.045, 0.060, respectively; *7*) fluid superheating zone for versions *1–3*; *8*) fluid superheating zone for versions *4–6*.

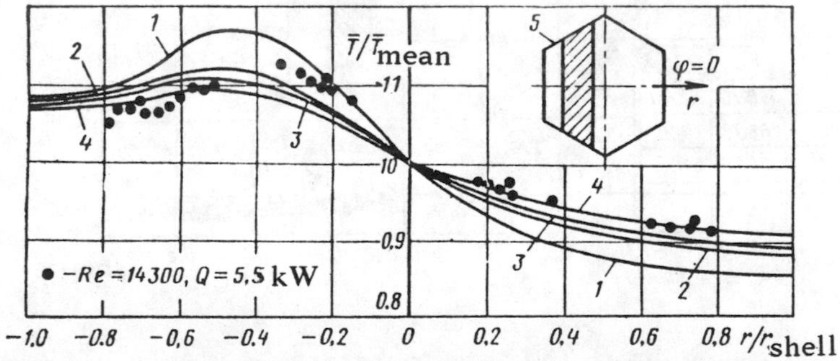

Figure 4.9 Fluid temperature fields for a helical tube bundle with Fr_M = 64 and an azimuthal direction φ = 0: *1–4*) calculation of Re = 14,300 and Q = 5.5 kW for coefficients k = 0.045, 0.095, 0.110, 0.145, respectively; *5*) location region of heated tubes in heat exchanger.

Let us consider in more detail the experimental and calculation results presented in Figs. 4.7–4.9. Experimentally measured temperatures of the heat carrier with azimuthal nonuniformity of heat release are in fair agreement with the predicted temperature curves mainly near the axis $\varphi = 0$, and a good deal of discrepancy of these temperatures is observed on the bundle periphery. According to tube shell wall calculations, the maximum heat carrier temperature should be equal to $\bar{T}$ = 452.8 K, and experimentally a temperature $\bar{T}$ = 390 K is observed. According to tube shell wall calculations, the minimum heat carrier temperature should be $\bar{T}$ = 280.8 K, but experimentally, $\bar{T} \approx$ 285 K. This discrepancy may be attributed to the rotational motion of the heat carrier near the tube shell wall relative to the tube bundle axis, which was also observed in [8]. This motion does not affect significantly either the coefficient k, determined with axisymmetric nonuniformity of heat release, or the coefficient k in the central region of the flow in a helical tube bundle with axisymmetric nonuniformity of heat release. In Fig. 4.7, where the effect of Re and heat release power Q on the temperature fields are considered, it is seen that over the investigated range of Re and Q, these equally influence the experimental and calculated temperature fields and do not influence the value of the coefficient k at a given Fr_M. The results on the effect of the length of the region where the heated tubes are located on the temperature fields (Fig. 4.8) show that, in this case, the experimental and calculated temperature fields are distorted equally at an invariable value of the dimensionless effective diffusion coefficient k. Figure 4.9 shows a comparison of the experimental and calculated temperature fields for a shape of the heated tube region which differs from the ones shown in Figs. 4.7–4.8. In this case, the impact of the rotational motion around the bundle periphery on the levelling of the developed temperature field nonuniformity is partly excluded. The experimental data on the coefficient k for azimuthal nonuniformity of heat release in helical tube bundles within the scatter of the experimental data practically coincide, at the same Fr_M number and Re > 10^4, with the coefficients k determined for the axisymmetric case.

Thus, the effective diffusion coefficients may be used to calculate three-dimensional velocity and temperature fields of the heat carrier in a helical tube heat exchanger with lateral heat carrier supply to the intertube space. The impact of rotational motion of the heat carrier near the heat exchanger casing relative to the bundle axis on the temperature fields in the peripheral cells with azmuthal nonuniformity of heat supply must be taken into account when estimating a heat exchanger temperature regime.

4.5 THE EFFECTIVE DIFFUSION COEFFICIENT AFFECTED BY MIXING PARAMETERS IN A HELICAL TUBE BUNDLE

The coefficient $\bar{k} = D_t/ud_{ef}$ determined by the method of diffusion from a system of linear sources, can be generalized by the relation

$$\bar{k}/k_{asy} = 1 - \exp\left[-0.0504(2ax/d)\right] \tag{4.63}$$

where the coefficients a and k_{asy} are determined by (4.55) and (4.47), respectively [12, 24, 25]. In this case, it is assumed that the Lagrange L_L and Euler L_E spatial integral turbulence scales are equal, the coefficient k depends only on Fr_M, and the bundle length $\bar{k} = k(Fr_M, x)$.

Use of Eq. (4.63) to generalize the experimental data on the coefficient $\bar{k}$ was justified within the framework of the assumptions made. However, the available experimental data make it possible to refine the effect of such parameters as the length of a helical tube bundle, inlet turbulence level, Reynolds number, and spatial integral turbulence scales on the coefficient k. Indeed, according to [54], on the circular tube axis the ratio $L_E/L_L \neq 1$, which may also be observed in a helical tube bundle, and an estimate of the entrance length allows the effect of the tube bundle length on the coefficient $\bar{k}$ to be assumed smaller than follows from Eq. (4.63). In [18], a study was made of flow mixing in helical tube bundles 0.5 and 1.5 m long with $Fr_M = 56$ and 220 over a Reynolds number range of $Re = 3.4 \cdot 10^3 - 3.8 \cdot 10^4$, porosity $m = 0.49 - 0.511$, and at an inlet turbulence level $\varepsilon = 1 - 6\%$. This study was made by the method of diffusion from linear sources on the experimental set-up described in § 2.4. The axisymmetric problem of levelling temperature nonuniformities was solved in the statement of the homogenized medium. Air served as the heat carrier. The maximum error in determining the coefficient k was 25-30%. Temperature fields were calculated when a nonlinear system of parabolic-type equations (3.8)-(3.14) was solved using the explicit network method [12]. The modified least-square and mathematical statistics methods were adopted to compare the experimental and calculated temperature fields and enabled determination of the most likely values of the dimensionless coefficient k and the likely ranges of the quantity k using experimental sampling.

The results of flow mixing are shown in Fig. 4.10, from which it follows that the Reynolds number affects the coefficient Fr_M only in the range of $Re < 10^4$ independent of the Fr_M number. The turbulence level, ε, at the bundle inlet also does not influence this coefficient. Since it is known that ε may exert an influence on the transfer characteristics only in the entrance length of the flow, whose length decreases with increasing ε [54], a need arises to estimate, using another method, the stabilized flow length, $x_{initial\ mixing}$, under conditions of axisymmetric nonuniformity of the energy release over a tube bundle radius. Equation (4.53) can be used to determine the length of the section on which a stabilized temperature or velocity profile of the heat carrier develops, when a jet having the same size as the maximum tube oval spreads in a tube bundle. The development of a flow core temeprature nonuniformity having the size of maximum oval, when the tubes are electrically heated, requires a length $x_{in}/d_{eq} = 14$, equal to the thermal entrance length when a tube bundle is uniformly heated over a bundle radius. Thus, the entrance length $x_{in.mix}$ at axisymmetric nonuniformity of energy release will be the sum of these two lengths and is determined by the relations [18]:

$$x_{in.mix}/d_{eq} = 8.019\ Fr_M^{0.226} \tag{4.64}$$

or

$$x_{in.mix}/d_{eq} = 12.1a^{-1}\ Fr_M^{-0.275} \tag{4.65}$$

Calculation of the length $x_{in.mix}$ by (4.64) and (4.65) shows that all helical tube bundles considered in the present study had a length $l > x_{in.mix}$. As a result, the assumption may be made that the coefficients k were investigated in the stabilized flow region, and the determined values of the coefficient k are the stabilized values of k_{stab}. Therefore, the scatter of the experimental data may be attributed to the accuracy in determining k, as well as to the effect of the bundle porosity. Indeed, the coefficients k for the same Fr_M number but for different bundle lengths were determined on different-porosity

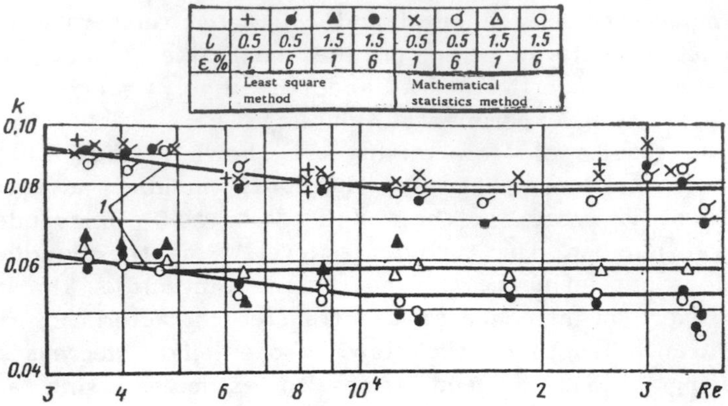

Figure 4.10 Effective diffusion coefficient versus Reynolds number, inlet turbulence level, and helical tube bundle length at $Fr_M = 56$: *1*) lines for the behavior of the diffusion coefficient as a function of Re

bundles because it was difficult to provide the same tube arrangement for different-length bundles. Hence, the bundle porosity with respect to the heat carrier may be considered as one of the characteristic parameters. Bauman's works also revealed the effect of the bundle porosity of finned rods on the mixing velocity. It appears that the mixing velocity of the heat carrier decreases with decreasing bundle porosity. Therefore, the criterial relation for k_{stab} will be sought in the form:

$$k_{stab} = k(Fr_M, Re, m) \tag{4.66}$$

Figure 4.11 presents the experimental results of different authors in the form of (4.66) for $Re \geq 10^4$. In this Reynolds number range, the coefficient k practically does not depend on Re (Fig. 4.10), and the experimental data are well represented by the relation

$$k_{stab} = 0.136 Fr_M^{-0.256} + 10 Fr_M^{-0.66} (m - 0.46) \tag{4.67}$$

The range of $Re = 3.4 \cdot 10^3 - 10^4$ is characterized by the effect of Re on the coefficient k_{stab} similar to that on the turbulent diffusion coefficient in a circular tube. Hence, the following formula

$$k_{stab} = 3.1623 \, [0.136 \, Fr_m^{-0.256} + 10 Fr_m^{-0.66} (m - 0.46)] Re^{-0.125} \tag{4.68}$$

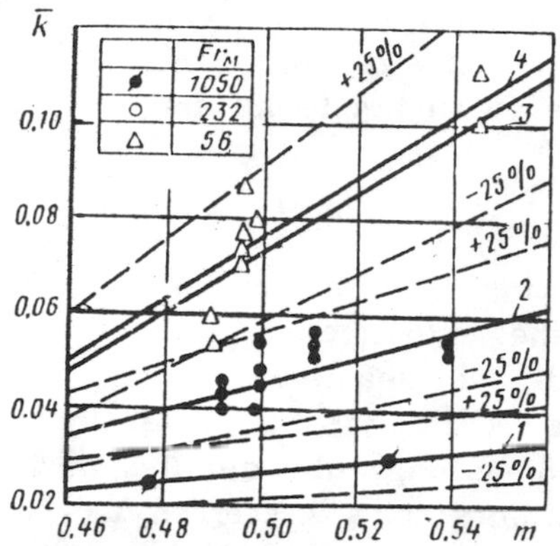

Figure 4.11 Mean effective diffusion coefficient versus porosity of helical tube bundle and Fr_M at Re > 10^4: *1–4*) Eq. (4.67) at Fr_M = 1050, 232, and 56, respectively

is valid. Equations (4.67) and (4.68) were obtained over the following parameter ranges: m = 0.46–0.56, Fr_M = 55–1080, ε = 1–6%, and $l/d_{eq} > x_{in,mix}/d_{eq}$, where $x_{in,mix}$ is determined by expressions (4.64) and (4.65).

The coefficients k_{stab} may be related to the coefficients k_{asy}. Considering that these coefficients are proportional to the spatial turbulence scales L_E and L_L, we have, respectively,

$$k_{stab}/k_{asy} = L_E/L_L \tag{4.69}$$

From Eq. (4.69) it is possible to obtain the order of magnitude of the ratio L_E/L_L. For bundles with m = 0.475 and Re = $8 \cdot 10^3$ (§ 4.2), the ratio L_E/L_L may be expressed as a function of Fr_M

$$L_E/L_L = 0.785\ Fr_M^{-0.127} \tag{4.70}$$

Calculation by Eq. (4.70) shows that over the range of Fr_M = 55–1080 the ratio L_E/L_L changes within 0.5–0.3. This is, by an order of magnitude, consistent with Mickelson's data [54] for almost uniform and isotropic turbulence in the flow core of a 0.2 m diameter tube axis at Re = $2 \cdot 10^5$–$6 \cdot 10^{10}$. Mickelson measured separately longitudinal correlation coefficients and Lagrange correlation coefficients and determined a mean value of the ratio L_E/L_L = 0.6 in the above Reynolds number range which linearly increases with growing pulsational velocity.

Figure 4.12 illustrates the experimental data of different authors which were processed in the form of the functional relation $\bar{k}/k_{asy} = f(2ax/d)$ and $\bar{k}/k_{stab} = \varphi(2ax/d)$, where these are compared with Eq. (4.63) and

$$\bar{k}/k_{stab} = 1 \tag{4.71}$$

respectively. Equation (4.71) is valid at $2a{\cdot}x/d > 24.4\ Fr_M^{-0.275}$. The scatter of the experimental data with respect to (4.71) does not exceed the maximum error in determining the coefficient $\bar{k}$ (Fig. 4.12). At the same time, the experimental data on helical 1.5 m long tube bundles deviate markedly from Eq. (4.63). This points to a preferred use of Eqs. (4.67) and (4.68) that allow the system of differential equations for the flow in helical tube bundles to be closed.

Thus, helical tube bundles are characterized by vigorous flow interchannel mixing due to spiral heat carrier swirling, thereby causing transverse velocity components, additional turbulization, and secondary flow circulation. The effective diffusion coefficients responsible for cross mixing in such devices exceed by 10 times the turbulent diffusion coefficient on a circular tube axis.

Let us estimate the contribution of turbulent diffusion transfer to the effective diffusion coefficient, which is, indeed, an integral characteristic of the transfer process in a helical

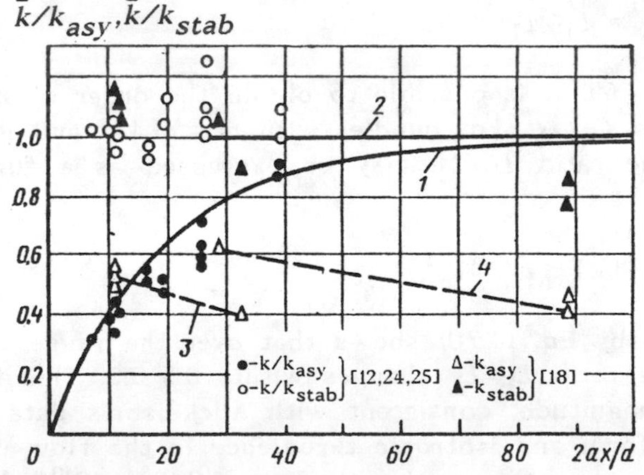

Figure 4.12 Comparison of different methods of generalizing experimental data: *1*) Eq. (4.63); *2*) Eq. (4.71); *3–4*) lines through the experimental points for bundles with Fr_M = 220 and 56, respectively.

tube bundle. For this, the experimentally determined turbulence intensity (3.54) and the integral turbulence scale (3.59) [16] can be used. Considering that the flow core turbulence is close to isotropic, and assuming that in a helical tube bundle the transverse integral turbulence scale $L_{2\ mean}$ is related to the scale $L_{1\ mean}$, as in the case of a flat channel, by $L_{2\ mean} = L_{1\ mean}/4$ [29], the turbulent diffusion contribution to cross mixing of the heat carrier may be estimated at Re = $8 \cdot 10^3$ by the relation:

$$\left(\frac{D_t}{u_{max} d_{eq}}\right)_{L_2} = \frac{\sqrt{\overline{u'^2}}}{u_{max}} \frac{L_{2\ mean}}{d_{eq}} = 0.0265\ \mathrm{Fr_M}^{-0.407} \left(1 + \frac{368}{\mathrm{Fr_M}}\right) \tag{4.72}$$

Since the quantities u' and L_1 are determined by the Euler flow description it is expedient to compare the quantity $(D_t/u_{max} d_{eq})L_2$ with the effective diffusion coefficients $D_t/u_{max}\ d_{eq}$ which, after a correction for scale difference in the Lagrange and Euler flow description is allowed in Eq. (4.47), may be expressed by the relation:

$$D_t/u_{max}\ d_{eq} = 0.0178\ (1 + 8.1\ \mathrm{Fr_M}^{-0.278}) \tag{4.73}$$

When calculating by Eq. (4.72), the relative contribution of turbulent diffusion to cross mixing of the heat carrier is estimated by the following quantities: 20, 13, and 5.3% of the quantity $D_t/u_{max}d_{eq}$ determined by (4.73) for bundles with $\mathrm{Fr_M}$ = 178, 296, and 1187, respectively. However, it should be noted that this is approximate since the assumption was made that in the flow core of a helical tube bundle the transverse turbulence scales are 4 times as small as the longitudinal ones, the same as in the case of a flat channel [29]. Indeed, according to the data of [15], the gradients of the transverse components of the average velocity are larger for bundles with a smaller value of $\mathrm{Fr_M}$, and the flow vorticity $\Omega = (\partial w/\partial y) - (\partial v/\partial z)$ will also increase with decreasing $\mathrm{Fr_M}$. This initiates vortex stretching in the longitudinal direction, increases L_1, and decreases L_2. Therefore, the contribution of turbulent transfer at small $\mathrm{Fr_M}$ may be somewhat less than follows from the above estimate.

CHAPTER

FIVE

HEAT TRANSFER AND HYDRAULIC RESISTANCE OF A CLOSE-PACKED HELICAL TUBE BUNDLE IN LONGITUDINAL FLOW

5.1 THEORETICAL MODELS FOR CALCULATING HEAT TRANSFER AND HYDRAULIC RESISTANCE

The model based on the semiempirical turbulence theories. The proposed model is a modification of the semiempirical Prandtl theory allowing for the specific features of swirling flow in helical tube bundles.

The physical model and design scheme of hydrodynamics. Let us consider a cell of a bundle having an infinite number of tubes (Fig. 5.1). For simplicity, this figure also shows the symmetric relative location of the tubes.

In the case of longitudinal flow past each tube, the fluid swirls, i.e., there appears a tangential velocity component w (Fig. 5.1). The relationship between the tangential and axial components depends on the relative tube twisting pitch s/d.

The flow is swirled by that part of the surface which is concave or convex relative to the axial flow, i.e., as shown in Fig. 5.2, it is mainly the surface d'. The concave part of the surface (in Fig. 5.2 it is the right side) initiates a normal component σ which is the wall response to the velocity head. The convex part of the surface produces a normal component by viscous forces. Thus, there appears a force moment acting from the wall side upon the flow and causing it to turn to the side of the surface twisting d'. Apparently, the remaining part of the tube surface (rounded end faces) contributes only insignificantly to flow swirling.

It is known [47, 56] that when a flow occurs in a tube with a spiral insert, forced vortex motion appears. In this case, the tangential velocity profile in the flow core, within a great

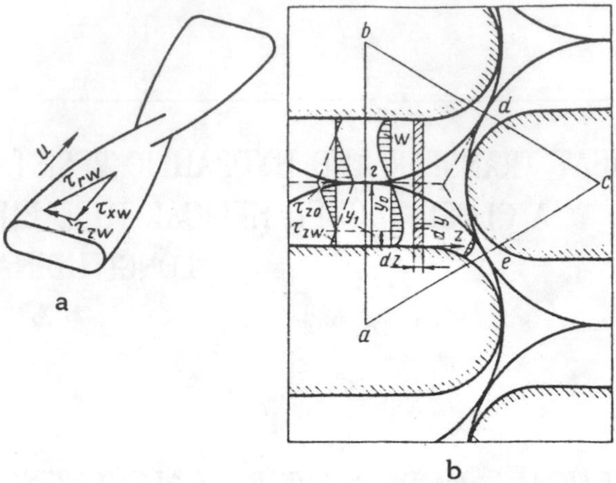

Figure 5.1 Physical flow model: *a)* shear stress components on the helical tube wall; *b)* cell of three helical tubes

accuracy, corresponds to the quasi-solid rotation law at a small distance from the spiral insert up to the wall layers of the tube. It is possible that in the case of flow past helical tube bundles, the flow pattern is of the same nature at some distance from the tube surface. However, closer to the middle of the flow, the flow pattern substantially changes. This is due to the fact that near adjacent tubes the vectors of a

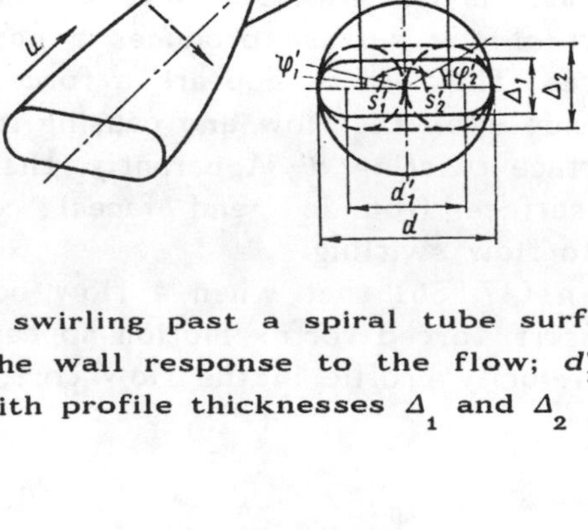

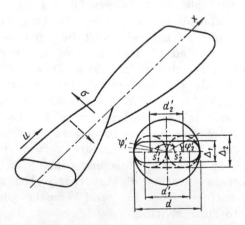

Figure 5.2 Flow swirling past a spiral tube surface: *σ)* normal component of the wall response to the flow; d'_1, d'_2) inducing tube surfaces with profile thicknesses Δ_1 and Δ_2

tangential velocity component are opposite at the same angular direction of tube twisting (Fig. 5.1). Thus, the tangential velocity profile far from the walls passes through a zero value. It is obvious that, in the simplest case, on the *ab* line the point at which $w = 0$ lies in the middle between the tubes, i.e., on the channel symmetry axis. For the remaining part of the flow, the location of points with $w = 0$ cannot be clearly identified. However, two limiting assumptions are possible: 1) tangential velocity components tend to zero on the lines of the circumscribed circles; 2) tangential velocity components tend to zero on lines close to the symmetry axes of the channel cross section. As shown below, both flow patterns cannot give substantially different calculation results on the hydraulic resistance and heat transfer.

The above specific nature of a tangential velocity profile initiates additional effects in the flow core as against the tube flow: i.e., tangential shear stresses (Fig. 5.1), turbulence generation, secondary flows, and Taylor's vortices. These effects result in increasing the hydraulic losses, increasing the heat transfer, and in levelling temperature nonuniformities of the heat carrier over the bundle cross section. At the same time, the main contribution to increasing the hydraulic resistance and the heat transfer is from the growing effective flow velocity, total shear stress, and generation of turbulent pulsations near the wall.

Comprehensive (theoretical) calculations of flow dynamics, including the determination of velocity fields and shear stresses in different zones of the cross section, postulate a knowledge of the perimeter distribution of such parameters as axial-to-tangential velocity component ratio and turbulent transfer coefficients. These data have recently been reported (Chapter 3) but are too few in number to perform complete calculations. Therefore, the present analysis has not aimed at determining local velocity profiles and shear stresses, but implying their subsequent integration over the entire cross section.

Note that in disjoined bundles of smooth tubes ($s_1/d > 1.5$, where s_1 is the tube location pitch), i.e., when the flow characteristics are distributed rather uniformly over the whole cross section, the real channel may be replaced by an equivalent ring. A complex flow pattern in twisted tube bundles impels consideration of an approximate flow model with the purpose of determining integral values of the hydraulic resistance and heat transfer coefficients. Assume that certain values of these coefficients are consistent with some mean effective degree of flow swirling, $\Gamma = \bar{w}/\bar{u}$, invariable along the tube perimeter. By flow swirling degree is understood the channel height-mean tangential-to-mean axial velocity component ratio at a given point on the perimeter.

Use of this approach is also justified by the fact that in a helical tube bundle the tube perimeter distribution of flow characteristics, namely, shear stress and the turbulent momentum transfer coefficient, is, to a lesser degree than in a smooth tube bundle with $s_1/d < 1.5$, set by the geometry of the channel cross section. This is favored by intensive secondary vortex exchange and by the absence of stagnation zones at the locations of contacting tubes. This is because the latter contact at points but not on lines, as in the case of a close-packed smooth tube bundle. Therefore, in a helical tube bundle the local flow characteristics must depend mainly on the degree of flow swirling.

Thus, consideration is given to the flow in an elementary cell with a bottom dz (Fig. 5.1) in which the flow has some height-mean values of tangential $\bar{w}$ and axial $\bar{u}$ velocity components. The quantity $\bar{u}$ defines the Reynolds number, and the quantity $\bar{w}/\bar{u}$ somewhat depends on the tube geometry and, in particular, on the tube twisting pitch s/d. In the middle of a given cell, when $y=y_0$, the tangential velocity w equals zero.

An equivalent design cell diameter is $d_{eq} = 4y_0$, i.e., it is determined in a fashion used to calculate flat channel flow when the flow parameters are invariable along the z-axis. In this case, the Reynolds number for a design cell is equal to Re for the entire abc channel (Fig. 5.1). The wall shear stress may also be resolved into two components: axial τ_{xw} and tangential τ_{zw}. The shear stress at the channel center consists only of a tangential component τ_{z0}.

As is well known, an analysis of the Navier-Stokes equations for steady fully developed turbulent incompressible flow in a flat channel and in a tube yields a linear shear stress distribution

$$\frac{d\tau_x}{dy} = \frac{dp}{dx} \tag{5.1}$$

Also, having the same relation for tangential motion, the law for a tangential shear stress component τ_z from τ_{zw} on the wall to τ_{z0} on the channel axis may be considered linear. In this case, at some distance from the wall y_1, where the tangential velocity achieves its maximum, w_{max}, the shear stress becomes equal to $\tau_z = 0$.

$$\tau_x = \tau_{xw}(1 - y/y_0) \tag{5.2}$$

$$\tau_z = \tau_{zw} + (\tau_{z0} - \tau_{zw}) \frac{y}{y_0} \tag{5.3}$$

where y_0 is the distance to the channel axis. According to the usual relations for turbulent flow, we have

$$\tau_x = (\mu + \mu_{tur\,x}) \frac{du}{dy} \tag{5.4}$$

$$\tau_z = (\mu + \mu_{tur\,z}) \frac{dw}{dy} \tag{5.5}$$

where μ_{turx} and μ_{turz} are the turbulent axial and tangential viscosity coefficients. Since the momentum of both velocity components is transferred by the same liquid particles, it may be assumed

$$\mu_{turx} = \mu_{turz} = \mu_{tur} \tag{5.6}$$

In what follows, use is made of the hypothesis proposed by Yu. A. Koshmarov [31] for rotational-translational liquid motion between two tangential cylinders rotating relative to one another. The turbulent viscosity at each point near a wall is assumed to depend on the specific features of the axial and tangential velocity distributions. Then, proceeding from the Prandtl hypothesis and having two limiting cases $\bar{u}=0$ and $\bar{w}=0$, the turbulent viscosity coefficient may be determined as:

$$\mu_{tur} = \rho l^2 \left\{\left|\frac{du}{dy}\right| + \left|\frac{dw}{dy}\right|\right\} \tag{5.7}$$

where l is the mixing length.

Following Prandtl's hypothesis, it is assumed that l, up to a certain distance from a wall, is governed by the relation:

$$l = \varkappa y \tag{5.8}$$

where $\varkappa$ is an empirical coefficient equal to 0.4 [55]. In [55] it is shown that Eq. (5.8) for the flow is valid only up to a distance from a wall equal to y/R = 0.2, where R is the tube radius. However, since in the Prandtl scheme l and the shear stress $\tau = \tau_W$ = const were simultaneously overestimated for the central flow portion, it became possible to obtain a velocity profile quite in agreement with experiment.

In the channel formed by a twisted tube bundle the central flow portion may be considered as consisting of intermittent jets. It is known [1] that in jet theory the mixing length in the mixing zone is assumed to be constant. In line with the above physical discussion, the following design scheme is proposed to determine l:

- in region 1, from the tube wall to the point y, where $w = w_{max}$

$$l = \varkappa y \tag{5.9}$$

- in region 2, from the point y to the channel axis,

$$l = \varkappa y_1 = \text{const} \tag{5.10}$$

In accordance with the two-layer Prandtl model, the flow is conventionally divided into a viscous sublayer, in which molecular transfer prevails, and a turbulent core, in which molecular viscosity may be neglected. The viscous sublayer thickness may be determined using the Prandtl hypothesis allowing for the flow swirl effect, i.e., in this case, the total wall shear stress τ_W must be considered as a characteristic quantity. Then, the viscous sublayer thickness

$$\delta_{VS} = \frac{\eta_V \nu}{\sqrt{\tau_{\Sigma w}/\rho}} \tag{5.11}$$

where $\tau_{\Sigma W} = \sqrt{\tau^2_{XW} + \tau^2_{ZW}}$, and η_V is an empirical constant equal to 11.5.

Since the viscous sublayer is thin, the velocity distribution in this layer may be assumed linear. Thus, a transformation of (5.4) and (5.5), taking account of (5.2), (5.3), and (5.7), yields a system of two nonlinear differential first-order equations

$$\tau_{xw}\left(1 - \frac{y}{y_0}\right) = -\left\{\mu + \rho l^2 \left[\left|\frac{du}{dy}\right| + \left|\frac{dw}{dy}\right|\right]\right\} \frac{du}{dy} \tag{5.12}$$

$$\tau_{zw}+(\tau_{z0}-\tau_{zw})\frac{y}{y_0}=-\left\{\mu+\rho l^2\left[\left|\frac{du}{dy}\right|+\left|\frac{dw}{dy}\right|\right]\right\}\frac{dw}{dy} \tag{5.13}$$

Solving these equations, one arrives at an expression for the axial and tangential velocity profiles in the design model as functions of τ_{xw}, τ_{zw}, τ_{z0}, which are the unknown quantities.

Determination of the axial and tangential velocity component profiles. In accord with the adopted design scheme, Eqs. (5.12) and (5.13) will be integrated with respect to three layers: 1) viscous sublayer; 2) turbulent core of region 1 from the viscous sublayer boundary to the point y_1, where $w = w_{max}$; 3) turbulent core of region 2 from the point y_1 to the channel axis.

The integration constants can be determined by considering the wall boundary conditions and the solution closure conditions for all three layers.

For convenience of further manipulations, these dimensionless quantities are defined:

$$z=\frac{\tau_{zw}}{\tau_{xw}};\quad z_0=\frac{\tau_{z0}}{\tau_{xw}} \tag{5.14}$$

$$\eta_x=\frac{y\sqrt{\tau_{xw}/\rho}}{\nu};\quad \eta_z=\frac{y\sqrt{\tau_{zw}/\rho}}{\nu} \tag{5.15}$$

$$\eta_{x0}=\frac{y_0\sqrt{\tau_{xw}/\rho}}{\nu};\quad \eta_{z0}=\frac{y_0\sqrt{\tau_{zw}/\rho}}{\nu} \tag{5.16}$$

$$\bar{\eta}=\frac{y}{y_0}=\frac{\eta_x}{\eta_{x0}}=\frac{\eta_z}{\eta_{z0}};\quad \bar{\eta}_1=\frac{y_1}{y_0} \tag{5.17}$$

$$\varphi_x=\frac{u}{\sqrt{\tau_{xw}/\rho}};\quad \bar{\varphi}_x=\frac{\bar{u}}{\sqrt{\tau_{xw}/\rho}} \tag{5.18}$$

$$\varphi_z=\frac{w}{\sqrt{\tau_{zw}/\rho}};\quad \bar{\varphi}_z=\frac{\bar{w}}{\sqrt{\tau_{zw}/\rho}} \tag{5.19}$$

$$\Gamma=\bar{w}/\bar{u};\quad \frac{\bar{\varphi}_z}{\bar{\varphi}_x}=\frac{\Gamma}{\sqrt{z}} \tag{5.20}$$

$$\mathrm{Re} = \bar{u}d_{eq}/\nu \tag{5.21}$$

$$\xi_x=\frac{\partial p}{\partial x}\cdot\frac{2d_{eq}}{\rho\bar{u}^2}=\frac{8\tau_{xw}}{\rho\bar{u}^2} \tag{5.22}$$

where $\bar{u}$ and $\bar{w}$ are the axial and tangential velocity components, respectively. The Reynolds number is determined in terms of the equivalent diameter and the axial velocity component. Also, let us introduce the following dimensionless quantities:

$$\eta_V = \frac{\delta_{V.S}\sqrt{\tau_{\Sigma w}/\rho}}{\nu} = \frac{\delta_{V.S}\sqrt{\tau_{xw}/\rho}}{\nu}(1+z^2)^{1/4} = \eta_{x_V}(1+z^2)^{1/4} \tag{5.23}$$

$$\eta_V = \frac{\delta_{V.S}\sqrt{\tau_{\Sigma w}/\rho}}{\nu} = \frac{\delta_{V.S}\sqrt{\tau_{zw}/\rho}}{\nu}\left(1+\frac{1}{z^2}\right)^{1/4} = \eta_{zV}\left(1+\frac{1}{z^2}\right)^{1/4} \tag{5.24}$$

$$\bar{\eta}_V = \frac{\eta_{xV}}{\eta_{x0}} = \frac{\eta_{zV}}{\eta_{z0}} \tag{5.25}$$

Proceeding from the linear profile, in the viscous sublayer we have:

$$\varphi_{x_V} = \eta_{x_V} = \frac{11.5}{(1+z^2)^{1/4}}\,; \quad \varphi_{z_V} = \eta_z = \frac{11.5}{\left(1+\frac{1}{z^2}\right)^{1/4}} \tag{5.26}$$

For region 1 of the turbulent core, Eqs. (5.12) and (5.13) may be written as:

$$\tau_{xw}\left(1-\frac{y}{y_0}\right) = \rho\varkappa^2 y^2\left(\frac{du}{dy}+\frac{dw}{dy}\right)\frac{du}{dy} \tag{5.27}$$

$$\tau_{zw}+(\tau_{z0}-\tau_{zw})\frac{y}{y_0} = \rho\varkappa^2 y^2\left(\frac{du}{dy}+\frac{dw}{dy}\right)\frac{dw}{dy} \tag{5.28}$$

As calculations have shown, the sizes of this region at the most are: y_1/y_0 = 0.2–0.26; therefore, Eq. (5.27) may be somewhat simplified, assuming $\tau_x = \tau_{xw}$ = const for this region.

For region 2 of the turbulent core, Eqs. (5.12) and (5.13) will assume the form:

$$\tau_{xw}\left(1-\frac{y}{y_0}\right) = \rho\varkappa^2 y_1^2\left(\frac{du}{dy}+\frac{dw}{dy}\right)\frac{du}{dx} \tag{5.29}$$

$$\tau_{zw}+(\tau_{z0}-\tau_{zw})\frac{y}{y_0} = \rho\varkappa^2 y_1^2\left(\frac{du}{dy}+\frac{dw}{dy}\right)\frac{dw}{dy} \tag{5.30}$$

Simultaneously solving Eqs. (5.27)–(5.30) with regard to (5.14) and (5.17)–(5.19), we have:

- for region 1 of the turbulent core

$$\frac{d\varphi_x}{d\bar{\eta}}=\frac{1}{\varkappa\bar{\eta}\sqrt{1+z+(z_0-z)\bar{\eta}}} \tag{5.31}$$

$$\frac{d\varphi_z}{d\bar{\eta}}=\frac{z+(z_0-z)\bar{\eta}}{\varkappa\bar{\eta}\sqrt{1+z+(z_0-z)\bar{\eta}}} \tag{5.32}$$

- for region 2 of the turbulent core

$$\frac{d\varphi_x}{d\bar{\eta}}=\frac{1-\bar{\eta}}{\varkappa\bar{\eta}_1\sqrt{1-z-(1+z_0-z)\bar{\eta}}} \tag{5.33}$$

$$\frac{d\varphi_z}{d\bar{\eta}}=\frac{z+(z_0-z)\bar{\eta}}{\varkappa\bar{\eta}_1\sqrt{1-z-(1+z_0-z)\bar{\eta}}} \tag{5.34}$$

Integration of Eqs. (5.31) and (5.32) from $\bar{\eta} = \bar{\eta}_V$ to $\bar{\eta} = \bar{\eta}_1$, and integration of Eqs. (5.33) and (5.34) from $\bar{\eta} = \bar{\eta}_1$ to $\bar{\eta} = 1$ yields the expressions for the axial and tangential velocity component profiles. Integration constants are determined from the conditions of closure for all three layers. As the thickness of a viscous sublayer is small, the approximate expression

$$\sqrt{1+z+(z_0+z)\frac{\eta_V}{(1+z^2)^{1/4}\eta_{x0}}}\approx 1+z+\frac{1}{2}(z_0-z)\frac{\eta_V}{(1+z^2)^{1/4}\eta_{x0}} \tag{5.35}$$

is valid.

Utilizing Eqs. (5.15), (5.16), (5.21), and (5.22), we obtain:

$$\eta_{x0}=\frac{\mathrm{Re}\sqrt{\xi_x}}{8\sqrt{2}}\,;\quad \eta_{z0}=\eta_{x0}\sqrt{z} \tag{5.36}$$

The coordinates of the tangential velocity maximum are found from Eqs. (5.32) and (5.34)

$$\bar{\eta}_1=-\frac{z}{z_0-z} \tag{5.37}$$

Upon integration of Eqs. (5.31)-(5.34), the formulas for the axial tangential velocity component profiles will assume the form:

- Region 1 ($\bar{\eta}_V \le \bar{\eta} \le \bar{\eta}_1$ or $\delta_{V.S} \le y \le y_1$):

$$\varphi_x=\frac{1}{\varkappa\sqrt{1+z}}\ln\left|\frac{\sqrt{1+z+(z_0-z)\bar{\eta}}-\sqrt{1+z}}{\sqrt{1+z+(z_0-z)\bar{\eta}}+\sqrt{1+z}}\right|+C_{x1} \quad (5.38)$$

$$\varphi_z=\frac{\sqrt{z}}{\varkappa\sqrt{1+z}}\ln\left[\frac{\sqrt{1+z+(z_0-z)\bar{\eta}}-\sqrt{1+z}}{\sqrt{1+z+(z_0-z)\bar{\eta}}+\sqrt{1+z}}\right]$$
$$+\frac{2}{\varkappa\sqrt{z}}\sqrt{1+z+(z_0-z)\bar{\eta}}+C_{z1} \quad (5.39)$$

- Region 2 ($\bar{\eta}_1 \le \bar{\eta} \le 1$ or $y_1 \le y \le y_0$):

$$\varphi_x=-\frac{2}{\varkappa\bar{\eta}_1(1+z_0-z)}\sqrt{1-z-(1+z_0-z)\bar{\eta}} \quad (5.40)$$
$$+\frac{2\sqrt{1-z-(1+z_0-z)\bar{\eta}}}{3\varkappa\bar{\eta}_1(1+z_0-z)^2}[(1+z_0-z)\bar{\eta}+2(1-z)]+C_{x2}$$

$$\varphi_z=-\frac{2\sqrt{z}}{\varkappa\bar{\eta}_1(1+z_0-z)}\sqrt{1-z+(1+z_0-z)\bar{\eta}}$$
$$-\frac{2(z_0-z)\sqrt{1+z-(1+z_0-z)\bar{\eta}}}{3\varkappa\bar{\eta}_1\sqrt{z}(1+z_0-z)^2}[(1+z_0-z)\bar{\eta}+2(1-z)]+C_{z2} \quad (5.41)$$

The integration constants are determined by the following expressions:

$$C_{x1}=\frac{\eta_V}{(1+z^2)^{1/4}}+\frac{1}{\varkappa\sqrt{1+z}}\ln\left[\frac{4}{\frac{z_0-z}{1+z}\,\frac{\eta_V}{(1+z^2)^{1/4}}\,\frac{8\sqrt{2}}{\mathrm{Re}\sqrt{\xi}}}+1\right]$$

$$C_{z1}=\frac{\eta_V}{(1+1/z^2)^{1/4}} \quad (5.42)$$

$$+\frac{\sqrt{z}}{\varkappa\sqrt{1+z}}\ln\left[\frac{4}{\frac{z_0-z}{1+z}\cdot\frac{\eta_V}{(1+1/z^2)^{1/4}}\cdot\frac{8\sqrt{2}}{\mathrm{Re}\sqrt{\xi}\sqrt{z}}}+1\right]$$

$$-2\sqrt{1+z}\sqrt{1+\frac{z_0-z}{1+z}\,\frac{\eta_V}{(1+1/z^2)^{1/4}}\cdot\frac{8\sqrt{2}}{\mathrm{Re}\sqrt{\xi}\sqrt{z}}} \quad (5.43)$$

$$C_{x2}=\frac{1}{\varkappa\sqrt{1+z}}\ln\left|\frac{1-\sqrt{1+z}}{1+\sqrt{1+z}}\right|+\frac{2}{\varkappa\bar{\eta}_1(1+z_0-z)}\sqrt{1+\frac{z}{z_0-z}}$$
$$+\frac{2\sqrt{1+\frac{z}{z_0-z}}}{3\varkappa\bar{\eta}_1(1+z_0-z)^2}\left[(1+z_0-z)\frac{z}{z_0-z}-2(1-z)\right]+C_{x1} \quad (5.44)$$

$$C_{z2}=\frac{\sqrt{z}}{\varkappa\sqrt{1+z}}\ln\left[\frac{1-\sqrt{1+z}}{1+\sqrt{1+z}}\right]+\frac{2}{\varkappa\sqrt{z}}+\frac{2\sqrt{z}}{\varkappa\bar{\eta}_1(1+z_0-z)}\sqrt{1+\frac{z}{z_0-z}}$$
$$-\frac{2(z_0-z)\sqrt{1+\dfrac{z}{z_0-z}}}{3\varkappa\bar{\eta}_1\sqrt{z}(1+z_0-z)^2}\left[(1+z_0-z)\frac{z}{z_0-z}-2(1-z)\right]+C_{z1} \qquad (5.45)$$

Thus, φ_X and φ_Z for both regions depend on z, z_0, ξ_X, η, and Re; in this case, z, z_0, and ξ_X depend on the degree of flow swirling and the Reynolds number.

Determination of the dimensionless shear stress components z and z_0. A system of three equations can be set up to determine ξ_X, z, and z_0 at a prescribed Reynolds number. Integration of Eqs. (5.38), (5.39), (5.42), and (5.43) with respect to the boundary layer thickness gives an expression for the mean axial and tangential velocity components:

$$\left.\begin{aligned}\bar{\varphi}_x&=\bar{\varphi}_{x1}\bar{\eta}_1+\bar{\varphi}_{x2}(1-\bar{\eta}_2)\\ \bar{\varphi}_z&=\bar{\varphi}_{z1}\bar{\eta}_1+\bar{\varphi}_{z2}(1-\bar{\eta}_2)\end{aligned}\right\} \qquad (5.46)$$

where

$$\left.\begin{aligned}\bar{\varphi}_{x1}&=\frac{1}{\bar{\eta}_1}\int_0^{\bar{\eta}_1} f_1(z,\ z_0,\ \xi_x,\ \bar{\eta},\ \mathrm{Re})\,d\bar{\eta}\\ \bar{\varphi}_{x2}&=\frac{1}{1-\bar{\eta}_1}\int_{\bar{\eta}_1}^{1} f_2(z,\ z_0,\ \xi_x,\ \bar{\eta},\ \mathrm{Re})\,d\bar{\eta}\\ \bar{\varphi}_{z1}&=\frac{1}{\bar{\eta}_1}\int_0^{\bar{\eta}_1} \psi_1(z,\ z_0,\ \xi_x,\ \bar{\eta},\ \mathrm{Re})\,d\bar{\eta}\\ \bar{\varphi}_{z2}&=\frac{1}{1-\bar{\eta}_1}\int_{\bar{\eta}_1}^{1} \psi_2(z,\ z_0,\ \xi_x,\ \bar{\eta},\ \mathrm{Re})\,d\bar{\eta}\end{aligned}\right\} \qquad (5.47)$$

Here, f_1 and f_2 are the relations for φ_X for regions 1 and 2 of the turbulent core according to Eqs. (5.38) and (5.40), and ψ_1 and ψ_2 are the appropriate relations for φ_Z by Eqs. (5.39) and (5.41).

At the same time, the mean velocity components may be found from Eqs. (5.20) and (5.22):

$$\bar{\varphi}_x=\sqrt{8/\xi_x};\quad \bar{\varphi}_z=\sqrt{8/\xi_x}\,\frac{\Gamma}{\sqrt{z}} \tag{5.48}$$

Also, using the boundary condition for the flow core, $\varphi_z = 0$ at $\bar{\eta} = 1$, a system of three transcendental implicit equations is obtained:

$$\left.\begin{aligned}&\sqrt{8/\xi_x}-\bar{\varphi}_{x1}\bar{\eta}_1+\bar{\varphi}_{x2}(1-\bar{\eta}_1)=0\\&\sqrt{8/\xi_x}\,\frac{\Gamma}{\sqrt{z}}-\bar{\varphi}_{z1}\bar{\eta}_1+\bar{\varphi}_{z2}(1-\bar{\eta}_1)=0\\&\psi_2(z,\,z_0,\,\xi_x,\,\mathrm{Re})_{\bar{\eta}_1=0}=0\end{aligned}\right\} \tag{5.49}$$

The system of equations (5.49) has been solved numerically; z, z_0, and ξ_x were determined at assigned values of Re and Γ.

Determination of the hydraulic loss coefficient. The balance of forces acting upon a fluid element $2y_0$ in height and $dxdz$ at the bottom (Fig. 5.1) is considered to determine the coefficient of total hydraulic losses. The equilibrium conditions in the axial and tangential directions may be written, respectively, as:

$$\frac{dp}{dx}\,dxdz2y_0=\tau_{xw}dxdz\cdot 2 \tag{5.50}$$

$$\frac{dp}{dz}\,dzdxy_0=\tau_{zw}dxdz+\tau_{z0}dxdz \tag{5.51}$$

Note that in Eq. (5.51), the surface over which the shear stress τ_{z0} acts is assumed to be equal to the wall where τ_{zw} and τ_{xw} take place. In a real channel these surfaces are different, to one degree or another, depending on at what place the flow core w becomes zero. For the considered tube geometry this difference is 30% if $w = 0$ on the symmetry axes and 25% if $w = 0$ on the lines of the circumscribed circles. In the calculation, the above difference in the surfaces is taken into account by an appropriate correction k. Then, from Eqs. (5.50) and (5.51) it is possible to determine the axial and tangential pressure losses

$$\Delta p_x=\frac{\tau_{xw}}{y_0}\,\Delta x \tag{5.52}$$

$$\Delta p_z=\frac{\tau_{zw}+\tau_{z0}k}{y_0}\,\Delta z=\frac{\tau_{zw}+\tau_{z0}k}{y_0}\,\Delta x\Gamma \tag{5.53}$$

The last expression is obtained by utilizing the relation:

$$\frac{\Delta z}{\Delta x}=\frac{w}{u}=\Gamma \qquad (5.54)$$

Total pressure losses are

$$\Delta p_{\Sigma}=\left(\frac{\tau_{xw}}{y_0}+\frac{\tau_{zw}+\tau_{z0}k}{y_0}\Gamma\right)\Delta x \qquad (5.55)$$

Defining the total hydraulic loss coefficient as

$$\xi=\left(\frac{dp}{dx}\right)_{\Sigma} d_{eq}\Big/\frac{\rho\bar{u}^2}{2} \qquad (5.56)$$

with regard to Eqs. (5.41) and (5.22), we arrive at:

$$\xi=\xi_x(1+z\Gamma+z_0\Gamma k) \qquad (5.57)$$

Determination of the relationship between the degree of flow swirling and the geometrical tube parameters. A solution to the system of equations (5.49) was obtained at a prescribed flow swirling degree Γ. In this case, it is necessary to specify for what helical tube geometry this value of Γ is appropriate.

Velocity profile calculations have shown that the values of the mean axial velocity component $\bar{u}$ are approximately consistent with the local axial velocity u at the point $y = y_1$. Therefore, the angular rotational velocity of flow particles with the coordinate $y = y_1$ is

$$\omega \doteq 2\pi\bar{u}/s \qquad (5.58)$$

In line with the above flow model, a change in the maximum tangential velocity, w_{max}, along the tube perimeter is assumed to be proportional to the radius r and the angular velocity: $w_{max} = \omega r$. Then, the mean-integral value may be found by

$$\frac{w_{max}}{\bar{u}}=\frac{1}{\pi/2}\int_0^{\pi/2}\frac{2\pi r}{s}\,d\varphi \qquad (5.59)$$

As calculations have shown, the mean-to-maximum tangential velocity ratio, $\bar{w}/w_{max}$, changes only slightly over a wide range of Γ and is 0.69–0.74. We obtain

$$\Gamma = K_1 \frac{d}{s} \tag{5.60}$$

where K_1 is the constant equal to 1.7. As shown below, the experimental data on the heat transfer and hydraulic resistance agree well with the predicted values when Eq. (5.60) is used over a wide range of s/d.

The strong relationship between Γ and s/d is of importance. This allows a determination of K_1 using only the experimental results on hydraulic resistance with one tube, without having to assume a tangential velocity component distribution. The proposed model can then be employed to calculate the heat transfer and hydraulic resistance over a wide range of s/d and Re.

The methods of determining heat transfer. The above data on the velocity fields and the turbulent transfer coefficients make it possible to determine the heat transfer in helical tube bundles.

Let us write the energy equation for steady-state flow, neglecting longitudinal heat conduction:

$$\frac{\partial}{\partial y}\left[(\lambda + \lambda_{tur})\frac{\partial T}{dy}\right] = c_p \rho U_\Sigma \frac{\partial T}{\partial l} \tag{5.61}$$

where $u_\Sigma = \sqrt{u^2 + w^2}$ is the total velocity vector and l is the coordinate in the total velocity vector direction.

Integration of Eq. (5.61) from y_0 to y, when the boundary condition $\partial T/\partial\, y = 0$ at $y = y_0$ is used, gives $\partial T/\partial y$. Integration of the expression for $\partial T/\partial y$ from $y = 0$ to y, when the boundary condition $T = T_w$ at $y = 0$ is utilized, yields:

$$T - T_w = \frac{c_p \rho}{\lambda} \int_0^y \frac{\int_{y_0}^{y} u_\Sigma \frac{\partial T}{\partial l}\, dy}{1 + \lambda_{tur}/\lambda} \tag{5.62}$$

Let us reduce Eq. (5.62) to a dimensionless form:

$$\theta = \frac{\mathrm{Pe}}{4} \int_0^{\bar\eta} \frac{\int_1^{\bar\eta} U_\Sigma \frac{\partial \theta}{\partial L}\, d\bar\eta}{1 + \psi \,\mathrm{Pr}\, \frac{\mu_{tur}}{\mu}}\, d\bar\eta \tag{5.63}$$

where

$$\theta=\frac{T_W-T}{T_W-\bar{T}};\quad U_\Sigma=\frac{u_\Sigma}{\bar{u}_\Sigma};\quad L=\frac{l}{y_0};\quad \text{Pe}=\frac{\bar{u}_\Sigma d_{eq}}{a} \tag{5.64}$$

Here, $\bar{T}$ is the mean flow temperature in a given cross section, and $\bar{u}_\Sigma$ is the mean velocity in the same cross section. The parameter ψ characterizes the relationship between heat and momentum transfer due to turbulent pulsations.

For the wall heat flux density q_W = const along the tube length we have:

$$\frac{\partial T}{\partial l}=\frac{q_W}{c_p\rho\bar{u}_\Sigma y_0}\quad \text{or}\quad \frac{\partial\vartheta}{\partial L}=-\frac{\text{Nu}}{\text{Pe}} \tag{5.65}$$

The final expression for a temperature profile becomes

$$\theta=-\frac{\text{Nu}}{4}\int_0^{\bar{\eta}}\frac{\int_1^{\eta}U_\Sigma d\bar{\eta}}{1+\psi\Pr\frac{\mu_{tur}}{\mu}}d\bar{\eta} \tag{5.66}$$

Substitution of (5.66) into the expression for a mean dimensionless temperature

$$\bar{\theta}=\int_0^1 U_\Sigma\theta d\bar{\eta}=1 \tag{5.67}$$

yields an integral relation for heat transfer which is similar to the Layon integral [32]:

$$\text{Nu}=\frac{4}{\int_0^1\frac{\left(\int_0^{\bar{\eta}}U_\Sigma d\bar{\eta}\right)^2}{1+\psi\Pr(\mu_{tur}/\mu)}d\bar{\eta}} \tag{5.68}$$

In the main region of the turbulent flow, the velocity differs slightly from the mean in the cross section, i.e., in the flow core $U_\Sigma\approx 1$. Such an approximation has been utilized for tube flow [32]. It is still more appropriate to apply it to the present case when the velocity profile is more full. This assumption allows one to reduce Eq. (5.68) to:

$$\mathrm{Nu} = \frac{4}{\int\limits_0^1 \frac{(\bar{\eta}-1)^2}{1+\psi \Pr (\mu_{tur}/\mu)} d\bar{\eta}} \tag{5.69}$$

In integrating (5.69) we assume that Pr ≈ 1 and, in this case, $\psi \approx 1$. The two-layer flow model (a viscous sublayer and a turbulent core divided into two regions) is adopted to find μ_{tur}/μ in terms of the velocity fields and shear stresses. For the first region of the turbulent core ($\bar{\eta}_V < \bar{\eta} < \bar{\eta}_1$)

$$\frac{1}{\mu_{tur}/\mu} = \frac{\partial \varphi_X}{\partial \bar{\eta}} \cdot \frac{1}{\eta_{X0}} \left[1 - \frac{\varphi_X}{\frac{\partial_X}{\partial \bar{\eta}}} \frac{\partial}{\partial \eta} \left(\frac{1}{\cos \gamma_1}\right) \cos \gamma_2\right] \tag{5.70}$$

where the expression in brackets characterizes the increase in the turbulent viscosity due to the additional slip of the fluid layers, which is caused by the vector rotation with increasing distance from the wall. Here, γ_1 = arctg (w/u) and γ_2 = arctg (τ_Z/τ_X). As calculations have shown, because of this factor, the heat transfer increases by 8% at the maximum flow swirl degree Γ = 0.41. In the second region of the turbulent core, the above effect may be neglected. To simplify the integration, in the first region the quantity in brackets (5.70) was assumed constant and equal to its value at the viscous sublayer boundary. The final expression is of the form:

$$\mathrm{Nu} = 4\left\{\left[\frac{\bar{\eta}_V^3}{3} - \bar{\eta}_V^2 + \bar{\eta}_V\right] + \frac{8\sqrt{2}}{\psi \Pr \mathrm{Re} \sqrt{\xi_x} \varkappa}\right.$$

$$\times \left[1 - \frac{\varphi_x}{\partial \varphi_x/\partial \bar{\eta}} \frac{\partial}{\partial \bar{\eta}} \left(\frac{1}{\cos \gamma_1}\right) \cos \gamma_2\right]_{\bar{\eta}=\bar{\eta}_V} \left[-\frac{2(2+3z)}{3(z_0-z)^2}\right.$$

$$- \frac{2\sqrt{1+z+(z_0-z)\bar{\eta}_V}}{3(z_0-z)^2} \cdot ((z_0-z)\bar{\eta}_V - 2(1+z))$$

$$- \frac{4}{z_0-z}\left(1-\sqrt{1+z+(z_0-z)\bar{\eta}_V}\right) + \frac{1}{\sqrt{1+z}}\left(\ln\left|\frac{1-\sqrt{1+z}}{1+\sqrt{1+z}}\right|\right.$$

$$\left.\left.\left. - \ln\left|\frac{\sqrt{1+z+(z_0-z)\bar{\eta}_V} - \sqrt{1+z}}{\sqrt{1+z+(z_0-z)\bar{\eta}_{hV}} + \sqrt{1+z}}\right|\right)\right] + \frac{8\sqrt{2}}{\psi \Pr \mathrm{Re} \sqrt{\xi_x} \varkappa \bar{\eta}_1}$$

$$\times\Bigg[-\frac{2\sqrt{-z_0}}{(1+z_0-z)^3}\left(-\frac{z_0^2}{5}+\frac{2z_0(1-z)}{3}-(1-z)^2+\frac{2\sqrt{1+\frac{z}{z_0-z}}}{(1+z_0-z)^2}\right.$$

$$\left.\times\left(\frac{\left(1+\frac{z}{z_0-z}\right)^2}{5}-\frac{2(1-z)\left(1+\frac{z}{z_0-z}\right)}{3}+(1-z^2)\right)\right.$$

$$+\frac{4}{3}\frac{\sqrt{-z_0}}{(1+z_0-z)^2}((1+z_0-z)+2(1-z))$$

$$+\frac{4}{3}\frac{\sqrt{1+\frac{z}{z_0-z}}}{(1+z_0-z)^2}\left(\frac{(1+z_0-z)z}{z_0-z}-2(1-z)\right)$$

$$\left.-\frac{2}{1+z_0-z}\left(\sqrt{-z_0}-\sqrt{1+\frac{z}{z_0-z}}\right)\Bigg]\right\}^{-1} \quad (5.71)$$

The expressions for the temperature fields were obtained by integration of Eq. (5.66) at a known Nusselt number. The results are presented in § 5.2.

The model based on the concept of a wall layer thickness. This theoretical model, based on the method of solving the boundary layer equation [23] and adopted to calculate the heat transfer and friction, assumes that the flow model is implemented in a bundle when the concept of a characteristic wall layer thickness δ [59] is valid. In this case, the laws of heat transfer and resistance are reduced to the form:

$$\alpha_m=\alpha_m(z,\ z_m,\ \mathrm{Pr}) \quad (5.72)$$

$$\alpha=\alpha(z) \quad (5.73)$$

where

$$z=\frac{\mathrm{Re}_\vartheta}{\alpha}=\frac{u}{\alpha\mu}\int_0^\delta \rho\frac{u}{\bar{u}}\left(1-\frac{u}{\bar{u}}\right)dy \quad (5.74)$$

$$z_m=\frac{\mathrm{Re}_\vartheta}{\alpha_m}=\frac{\bar{u}}{\mu\alpha_m}\int_0^\delta \rho\frac{u}{\bar{u}}\,\frac{T-\bar{T}}{T_W-\bar{T}}\,dy \quad (5.75)$$

The subscript ϑ denotes that the Reynolds number is determined in terms of the momentum thickness, and the subscript ϑ means that the Reynolds number is found in terms of the energy loss thickness. The coefficients α and α_m are determined by Eqs. (3.32) and (3.43).

The velocity and temperature at the external boundary of a wall layer are taken as characteristic velocities and temperatures. When Eqs. (5.72) and (5.73) are used, it is possible to adopt boundary layer methods for calculating heat transfer and resistance at the entrance length of a helical tube bundle. These relations may be obtained as follows. The resistance law (3.33) and the logarithmic velocity-distribution law (3.31) are employed to obtain the resistance law in the form of (5.73). Then, from Eq. (5.74) we have:

$$z=\frac{\overline{\mathrm{Re}}_{\delta}}{0{,}39\sqrt{\bar{a}}}\left(1-\frac{2\sqrt{\bar{a}}}{0{,}39}\right) \tag{5.76}$$

where

$$\overline{\mathrm{Re}}_{\delta} = \frac{\rho\bar{u}\delta}{\mu} \approx 1.035\ \mathrm{Re}_{\delta} = 1.035\ \frac{\rho u_{\mathrm{mean}}\delta}{\mu} \tag{5.77}$$

or allowing for (3.38) in a more general form:

$$z=\frac{\overline{\mathrm{Re}}_{\delta}}{0{,}39\sqrt{\bar{a}}}\ \frac{1-\dfrac{2\sqrt{\bar{a}}}{0{,}39}}{1-\dfrac{4\delta^*}{d_{eq}}} \tag{5.78}$$

where δ^* is the displacement thickness.

$$\delta^*=\int_0^{\delta}\left(1-\frac{\rho u}{\bar{\rho}\bar{u}}\right)dy=\frac{z\alpha\mu_m}{\bar{u}\bar{\rho}}(H-H_{\rho}) \tag{5.79}$$

$$H=\frac{\int_0^{\delta}\rho\left(1-\frac{u}{\bar{u}}\right)dy}{\int_0^{\delta}\rho\frac{u}{\bar{u}}\left(1-\frac{u}{\bar{u}}\right)dy} \tag{5.80}$$

$$H_{\rho}=\frac{\int_0^{\delta}\rho\left(1-\frac{\bar{\rho}}{\rho}\right)dy}{\int_0^{\delta}\rho\frac{u}{\bar{u}}\left(1-\frac{u}{\bar{u}}\right)dy} \tag{5.81}$$

Considering the relationship between shear stress and resistance coefficient (3.36) and Eqs. (3.32) and (3.38), we have:

$$\xi=\frac{8\alpha}{\left(1-\frac{4\delta^*}{d_{eq}}\right)^2} \tag{5.82}$$

or at $u_{mean} \approx 0.95u$:

$$\xi=8{,}8\alpha \tag{5.83}$$

Substitution of $\overline{\mathrm{Re}}_\delta$ from (5.76) into (5.77) gives:

$$\overline{\mathrm{Re}}_\delta=0{,}39\,\frac{z\sqrt{\alpha}}{1{,}035\left(1-\frac{2\sqrt{\alpha}}{0{,}39}\right)} \tag{5.84}$$

or elimination of ξ from (5.82) and (5.83) gives:

$$\mathrm{Re}_\delta=\frac{0{,}266^4}{8{,}8^4\cdot\alpha^4} \tag{5.85}$$

Substitution of Re_δ into (5.74) and (5.78) reduces the resistance law for helical bundles to the form:

$$z=\frac{0{,}266^4\left(1-\frac{2\sqrt{\alpha}}{0{,}39}\right)1.035}{0{,}39\cdot 8{,}8^4\cdot\alpha^4\cdot\sqrt{\alpha}} \tag{5.86}$$

Given several values of α, Eq. (5.86) can be employed to make calculations. As a result, several values of z corresponding to given values of α are obtained. An approximation of the calculation results simplifies the resistance law for helical tube bundles to the form:

$$\alpha=0{,}045z^{-0{,}221}+4\cdot 10^{-4} \tag{5.87}$$

As shown below, the resistance law (5.87) may be utilized for the general case of fluid flow with variable physical properties. In this case, in Eq. (5.74) the viscosity coefficient μ and the liquid density ρ are determined at the mean mass liquid temperature T_{mean}.

Since the logarithmic temperature distribution (3.42) is satisfied over a range of parameters for the wall layer of a helical tube bundle, the quantity z_m will be:

$$z_m = \frac{\overline{\mathrm{Re}}_\delta}{0.39\sqrt{\bar{a}}}\left(1 - \frac{2\sqrt{\bar{a}}}{0.39}\right) = z \tag{5.88}$$

and instead of (5.72) we have:

$$\alpha_m = \alpha_m(z, \mathrm{Pr}) \tag{5.89}$$

Under conditions close to isothermal, the heat transfer law may be obtained in the following way. Introducing $q_w = \alpha_m \rho \bar{u} c_p t_{mean}$ into the formula for $\mathrm{Nu}_\delta = q_w \delta / t_{mean} \lambda$ and assuming $\bar{T} \approx T_{mean}$:

$$\mathrm{Nu}_\delta = \alpha_m \mathrm{Pr}\, \overline{\mathrm{Re}}_\delta \tag{5.90}$$

Substitution of Re_δ from (5.77) into (5.90) expresses Nu_δ as a function of α_m, z, Pr, and α:

$$\mathrm{Nu}_\delta = 0.39 z \alpha_m \mathrm{Pr} \frac{\sqrt{\bar{a}}}{\left(1 - \frac{2\sqrt{\bar{a}}}{0.39}\right)} \tag{5.91}$$

The heat transfer law may be given as

$$\mathrm{Nu}_\delta = 0.020\,\mathrm{Re}_\delta^{0.8} \cdot \mathrm{Pr}^{0.4} \tag{5.92}$$

Substitution of the expressions for Nu_δ and Re_δ from (5.91) and (5.85) into (5.92) leads to:

$$\alpha_m = 0.0195 z^{-0.2} \mathrm{Pr}^{-0.6} \left(\frac{0.39\sqrt{\bar{a}}}{1 - \frac{2\sqrt{\bar{a}}}{0.39}}\right)^{-0.2} \tag{5.93}$$

For various values of z, values of α can be determined by Eq. (5.87), and then the value of α_m by Eq. (5.93). As a result of this calculation, several values of α_m will be obtained. An approximation of these calculation results reduces the heat transfer law for helical tube bundles in nonisothermal flow to the following form:

$$\alpha_m = (30.4 z_1^{0.174} \mathrm{Pr}_m^{0.6} + 14.65 z_1^{0.03} - 11.2)^{-1} \tag{5.94}$$

where

$$z_1 = 1.035\,\mathrm{Re}_{\delta m} \frac{\left(1 - \frac{2\sqrt{\bar{a}}}{0.39}\right)}{0.39\sqrt{\bar{a}}} \tag{5.95}$$

$$\alpha_m = \frac{\mathrm{Nu}_{\delta m}}{0.965\,\mathrm{Re}_{\delta m}\,\mathrm{Pr}_m} \tag{5.96}$$

Laws of resistance (5.87) and heat transfer (5.94) for a helical tube bundle may be employed to calculate the friction and heat transfer in the entrance length by the boundary layer methods.

In [23], the differential boundary layer equations were simplified to the integral momentum and energy relations, whose solution allows a determination of the unknown arguments z and z_m.

The integral momentum relation

$$\frac{d}{dx}\left[\bar{u}^2 \int_0^\infty \rho \frac{u}{\bar{u}}\left(1-\frac{u}{\bar{u}}\right) dy\right] + \bar{u}\frac{d\bar{u}}{dx}\int_0^\infty \rho\left(\frac{\bar{\rho}}{\rho}-\frac{u}{\bar{u}}\right) dy = \tau_w \tag{5.97}$$

and the integral energy relation

$$\frac{d(\bar{u}\bar{j}_0)}{dx}\int_0^\infty \rho\frac{u}{\bar{u}}\left(\frac{J-\bar{J}}{J_w}\right) dy = q_w \tag{5.98}$$

were derived under the following boundary conditions:

$$\left.\begin{aligned} &y=0;\ u=0;\ v=0;\ \tau=\tau_w;\ q=q_w \\ &y=\infty\,(y=\delta);\ u=\bar{u};\ \tau=0;\ j=\bar{j};\ q=0 \end{aligned}\right\} \tag{5.99}$$

In [23], equations (5.97) and (5.98) are reduced, respectively, to the form:

$$\frac{dz}{d\,\mathrm{Re}_x} + \frac{z}{\alpha}\frac{d\alpha}{d\,\mathrm{Re}_x} + k\frac{d\bar{u}}{d\,\mathrm{Re}_x} = 1 \tag{5.100}$$

$$\frac{dz}{d\,\mathrm{Re}_x} + \frac{z_m}{\alpha_m}\frac{d\alpha_m}{d\,\mathrm{Re}_x} + \frac{z_m}{\bar{j}_0}\frac{d\bar{j}_0}{d\,\mathrm{Re}_x} = 1 \tag{5.101}$$

$$d\,\mathrm{Re}_x = \frac{\rho_x \bar{u}\,dx}{\mu_x} \tag{5.102}$$

$$k = 1 + H - H_\rho \tag{5.103}$$

Equations (5.100) and (5.101) were solved under the following assumptions [23]. Friction, heat transfer, and the velocity and temperature profiles in each cross section x of the boundary layer are specified by flow conditions, velocity, and temperature at the external boundary of the layer, by wall temperature, longitudinal pressure gradient (or $d\bar{u}/dx$), and by some two characteristics of the developed boundary layer. The integrals included in determination of z and z_m, (5.74) and (5.75), are taken as characteristics of the boundary layer. These integrals are equal to the product of some mean liquid density values over the boundary layer into the momentum and energy loss thickness, respectively.

In the stabilized flow length, where the characteristic wall layer thickness δ = const (x), the heat transfer and resistance laws in the form of (5.94) and (5.87), respectively, are also satisfied.

5.2 CALCULATION OF VELOCITY FIELDS, SHEAR STRESSES, COEFFICIENTS OF HYDRAULIC RESISTANCE, AND HEAT TRANSFER

Calculation results of velocity fields and hydraulic resistance. Solving the system of equations (5.49), one arrives at the values of τ_{xw}, τ_{zw} and τ_{z0} in a dimensionless form in terms of ξ_x z_0, z as a function of Re and Γ. At $\Gamma = 0$ the values of ξ_0 coincide with those calculated by the Blasius formula for a tube. Flow swirling increases the value of ξ_x. Axial velocity component losses increase by 24% with an increase of Γ from 0 to 0.41. Tangential shear stresses, z, on the wall and in the flow core are approximately proportional to the degree of flow swirling Γ; z_0 being approximately 4 times as large as that of z. The direction of the total shear stress vector, τ_{zw}, is consistent with the generatrix vector of the twisted surface of the tube (Fig. 5.1).

Figure 5.3 illustrates axial and tangential velocity component fields at several values of Γ. With increasing flow swirling (with decreasing s/d), the velocity profile becomes more full due to the increasing turbulent viscosity. In this case, the viscous sublayer thickness decreases. At Γ = 0.41 (s/d = 4.15) the viscous sublayer thickness is 28% less than in the nonswirled case. At Γ = 0 ($s/d \to \infty$) the calculated axial velocity profile coincides with the logarithmic one:

$$\varphi_x = 5.75 \lg \eta + 5.5 \tag{5.104}$$

The tangential velocity component (Fig. 5.3) has a maximum at $\bar{\eta}_1$ = 0.2–0.27 (when Γ varies from 0.41 to 0). The direction of the total velocity vector u_Σ determined by the ratio w/u changes over the channel height. Near the wall it

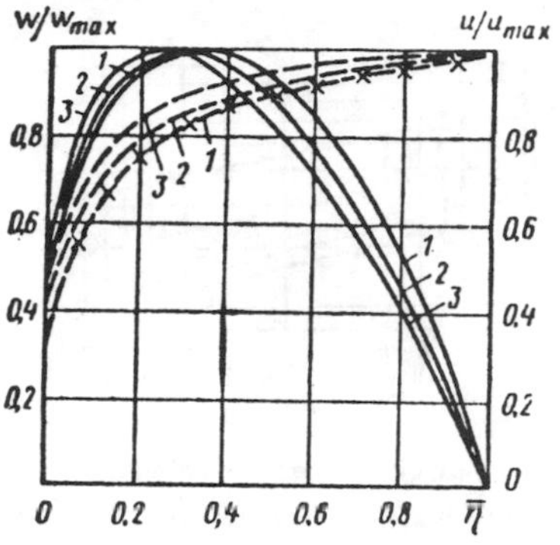

Figure 5.3 Profiles of dimensionless axial (dash) and tangential (solid lines) velocity components at Re = 10^4: *1, 2, 3*) $\bar{w}/\bar{u}$ = 0.0395, 0.135, 0.41, respectively; *x*) universal velocity profile for a tube

agrees approximately with the total shear stress direction. With increasing distance from the wall, the vector u_Σ turns to the side of the axial direction, approaching it at the channel center.

As seen from Figs. 5.4 and 5.5, the coefficient of the total hydraulic losses depends substantially on the twisting pitch s/d. At $s/d \to \infty$ the rlation is consistent with the Blasius formula. The most substantial increase in hydraulic losses due to flow swirling is observed at $s/d < 8$–10. Thus, at $s/d = 4.15$ the hydraulic losses exceed by 2.9–3.3 times those in the nonswirled flow, ξ_0. ξ as a function of s/d is well described by the available experimental data on helical tube bundles and on helically finned tube bundles.

The calculations allow an estimation of the hydraulic loss components, which are of importance to understanding mixing and heat transfer augmentation. Figure 5.6 shows the total hydraulic loss coefficients ξ and the coefficients ξ_W allowing for wall friction losses alone:

$$\xi_W = \xi_X (1 + z\Gamma) \tag{5.105}$$

It is obvious that $(\xi - \xi_W)$ characterizes the value of the hydraulic losses in the flow core, by which are understood the losses due to the shear stress τ_{z0} in the central channel part where the tangential velocity component undergoes a discontinuity. At $s/d > 12$ the flow core losses are essentially

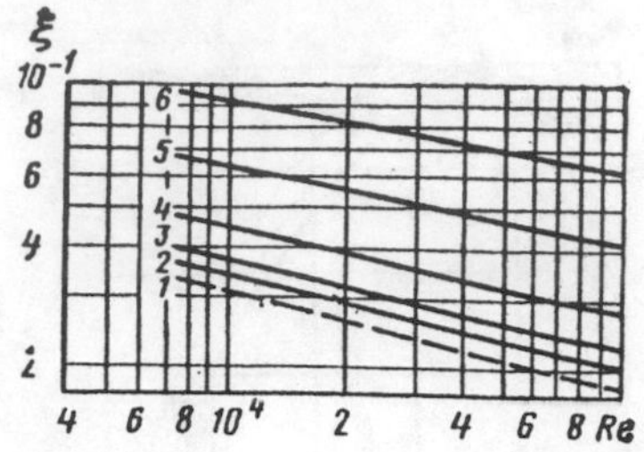

Figure 5.4 Friction factors at different $\Gamma = \bar{w}/\bar{u}$ (and at different s/d: *1)* $\Gamma = 0$ ($s/d \to \infty$); *2)* $\Gamma = 0.887$ ($s/d = 19.1$); *3)* $\Gamma = 0.135$ ($s/d = 12.45$); *4)* $\Gamma = 0.205$ ($s/d = 8.3$); *5)* $\Gamma = 0.307$ ($s/d = 5.54$); *6)* $\Gamma = 0.41$ ($s/d = 4.15$)

negligible and amount to only 8% at $s/d = 12$. These losses sharply increase with decreasing s/d and amount to 50–55% of the total losses at $s/d = 4.15$. As previously mentioned, in the central channel portion, the flow behaves as intersecting jets. The larger the shear stress, the higher is the turbulence generation near the jet interface. The turbulence intensity

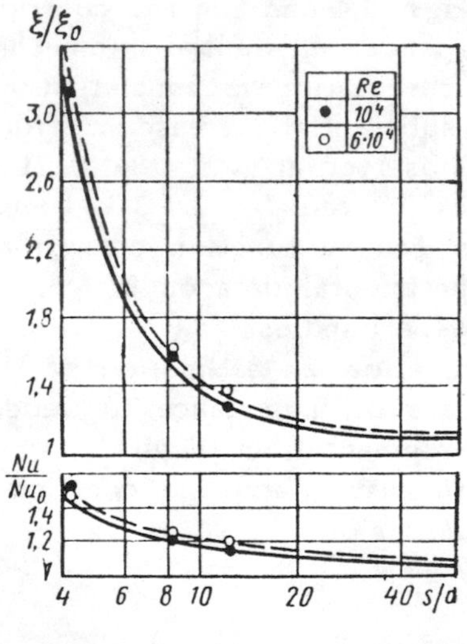

Figure 5.5 Friction factors and Nusselt numbers versus tube twisting pitch s/d: ξ_0 and Nu_0 are the friction factor and Nusselt number for nonswirled flow (at $\Gamma = 0$, $s/d \to \infty$). ——— and ---- represent theoretical calculation at Re = 10^4 and Re = $6 \cdot 10^4$, respectively

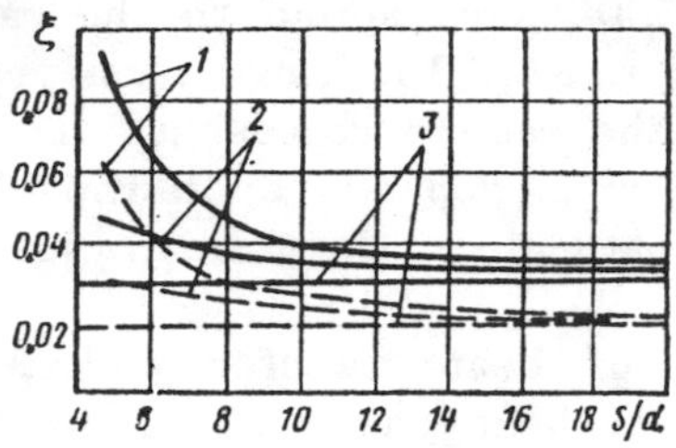

Figure 5.6 Hydraulic resistance coefficient components versus s/d at Re = 10^4 (solid lines) and Re = $8 \cdot 10^4$ (dash): *1)* total hydraulic resistance coefficients ξ; *2)* wall shear stress coefficient ξ_w; *3)* resistance coefficient ξ_0 in nonswirled flow

increases at the boundary of a jet and a fixed flow. It is higher in the case of a swirled jet as compared to a nonswirled one. Therefore, the mixing processes are greatly enhanced at $s/d < 12$.

The calculation results were approximated by an equation of the form:

$$\xi = f(\Gamma, \mathrm{Re}) \tag{5.106}$$

since z and z_0 at a given Re depend in a unique fashion on Γ. In turn, Γ at a fixed Re depends on s/d and the shape of the tube cross section. In line with the proposed model, the fuller the profile of the tube cross section (which means a smaller equivalent cell diameter, d_{eq}, Fig. 5.2), the smaller the flow swirling created by the tube at a given helical tube pitch. These physical representations are satisfied with the functional relationship:

$$\Gamma = f\left(\frac{1}{\frac{s}{d} \cdot \frac{s}{d_{eq}}}\right) \tag{5.107}$$

The calculation results for Re = $8 \cdot 10^3$–10^5 and $s/d > 4$ are, with a maximum deviation of ±3%, approximated by the relation:

$$\xi = \frac{0,25}{\mathrm{Re}^{0,22}}\left[\left(1 + \frac{\pi^2}{0.9\frac{s}{d} \cdot \frac{s}{d_{eq}}}\right)^{1,5} + \frac{100}{\left(\frac{s}{d} \cdot \frac{s}{d_{eq}}\right)^{1,25}}\right] \tag{5.108}$$

The terms of Eq. (5.108) correspond to the calculated hydraulic loss components (Fig. 5.6). The first term represents the wall friction losses and the second represents the flow core losses. Note that the first term depends upon the increasing effective velocity and liquid path along the tube walls.

Calculation results of heat transfer and temperature fields. Calculation results of heat transfer are presented in Figs. 5.5 and 5.7. The relations are of the form:

$$\mathrm{Nu} = C \cdot \mathrm{Re}^{0.75} \tag{5.109}$$

where C depends on the degree of flow swirling. A substantial increase in heat transfer, as compared to the nonswirled flow case, is observed at $s/d < 8$–10 and amounts to 47% at $s/d = 4.15$. The total thermal resistance is represented by the denominator of Eq. (5.71), and the thermal resistance of the viscous sublayer by the first term. Over the entire range of Γ, the thermal resistance of the viscous sublayer is 64%; at Re = 10^4 and 49% at Re = $8 \cdot 10^4$. Accordingly, the contribution to heat transfer augmentation due to decreasing the viscous sublayer thickness is 69% at Re = 10^4 and 44% at Re = $8 \cdot 10^4$. Thus, at Re = 10^4 the decrease in the viscous sublayer thickness due to increasing the total wall shear stress is of primary significance in improving the heat transfer. At higher Re $\approx 10^5$ the heat transfer improvement due to flow swirling, approximately to the same degree, occurs through decreasing the thermal resistance of the viscous sublayer and the turbulent core, in which the intensity of turbulent pulsations grows.

Similar to the velocity fields, the dimensionless temperature profile (Fig. 5.8) becomes more full with increasing flow swirling.

The calculated heat transfer data can be approximated, with a maximum deviation of ±2%, by the formula:

$$\mathrm{Nu} = 0.035\,\mathrm{Re}^{0.75}\left[1 + \frac{\pi^2}{0.5\dfrac{s}{d}\cdot\dfrac{s}{d_{eq}}}\right]^{0.4}\left[1 + \frac{1.3}{\left(\dfrac{s}{d}\cdot\dfrac{s}{d_{eq}}\right)^{0.6}}\right] \tag{5.10}$$

which is valid at Re = $6 \cdot 10^3$–10^5 and $s/d > 4$. Equation (5.110), in its structure, displays the above specific features of heat transfer enhancement. The term in the first bracket represents the heat transfer improvement with flow swirling due to decreasing the thermal resistance of the viscous sublayer, and

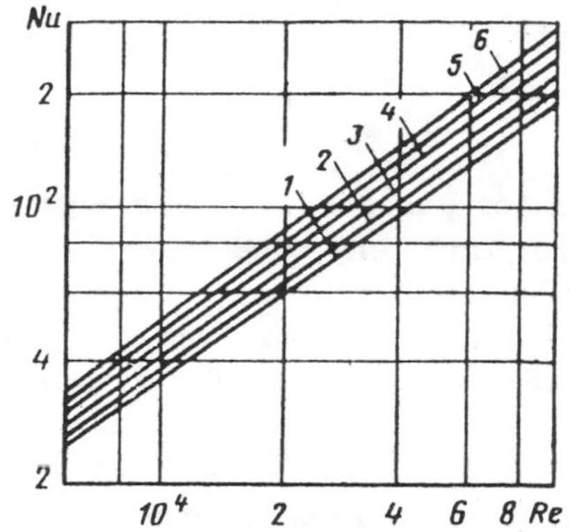

Figure 5.7 Nusselt numbers at different degrees of flow swirl Γ (at different s/d): *1)* $\Gamma = 0$ ($s/d \to \infty$); *2)* $\Gamma = 0.887$ ($s/d = 19.1$); *3)* $\Gamma = 0.135$ ($s/d = 12.45$); *4)* $\Gamma = 0.205$ ($s/d = 8.3$); *5)* $\Gamma = 0.307$ ($s/d = 5.54$); *6)* $\Gamma = 0.41$ ($s/d = 4.15$)

in the second bracket due to decreasing that of the turbulent core. In the limiting case, at $s/d \to \infty$, Eq. (5.110) practically coincides with the theoretical equation of S. S. Kutateladze [32]:

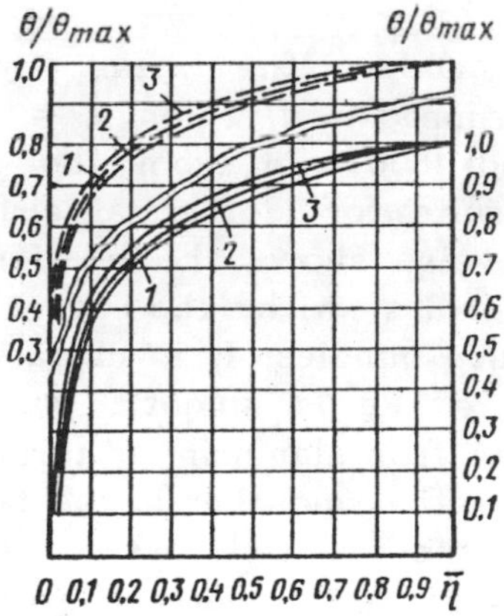

Figure 5.8 Dimensionless temperature profiles at different degrees of flow swirl Γ when Re = 10^4 (solid lines) and Re = $8 \cdot 10^4$ (dash): *1, 2, 3)* Γ = 0.0395, 0.135, and 0.41, respectively

$$\mathrm{Nu}=\frac{0.14\sqrt{\xi}\,\mathrm{Pr}\cdot\mathrm{Re}}{\ln\frac{\mathrm{Re}\sqrt{\xi}}{290}+4.6\,\mathrm{Pr}} \tag{5.111}$$

which was derived for gases and liquids with Pr < 5 using the two-layer model for turbulent tube flow.

The calculation results refer to gases with Pr ≈ 0.7. However, this calculation method may be used to calculate heat transfer to other fluids with Pr differing somewhat from 1.

Our calculations show that it is advisable to experimentally study heat transfer and hydraulic resistance in helical tube bundles in the region $s/d < 12$, where the flow swirling effects on Nu and ξ are most significant.

5.3 EXPERIMENTAL RESULTS OF HYDRAULIC RESISTANCE COEFFICIENTS

Results of hydraulic resistance coefficients for bundles composed of 19 helical tubes. A study was made at Re = 10^3–$8\cdot10^4$, i.e., it covered the developed turbulent flow regime, transition region, and laminar flow regimes with secondary vortices. Such an extension of the Reynolds number range facilitated an understanding of the flow pattern in helical tube bundles.

As mentioned above, the mean hydraulic resistance coefficient was determined in a region, l_0 = 500 mm long, at a distance of 200 and 1500 mm from the entrance and exit, respectively. The results for isothermal airflow are shown in Fig. 5.9. This figure also shows the relations for longitudinal flow past tube bundles with the same porosity as the considered helical tube bundles. It is characteristic that a plot of ξ as a function of Re is smooth, without a particularly pronounced transition from laminar to turbulent flow. In this case, the smaller s/d, the smoother is the plot of ξ versus Re.

The presence of the channel cross-section-uniform field of centrifugal forces, which are larger the greater is the distance from the tube center, initiates secondary flows (macrovortices or Taylor's vortices) and also favors the generation of turbulent vortices.

It may be assumed that, as in the case of flow in spiral tubes and in tubes with swirlers [56], the flow past helical

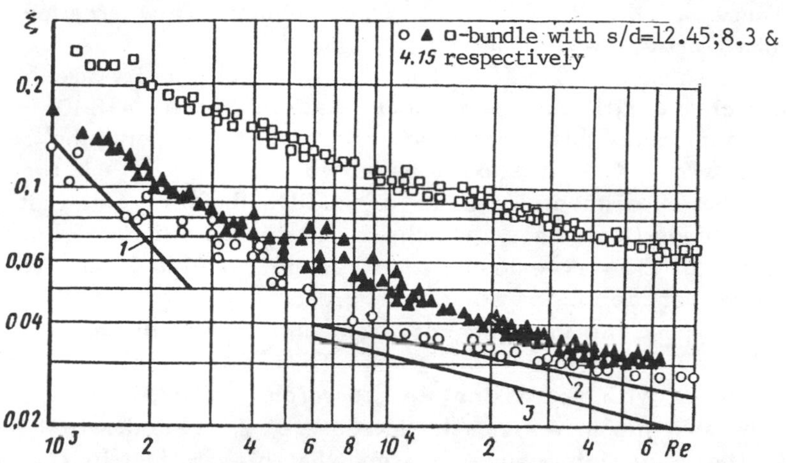

Figure 5.9 Friction factors in isothermal flow past a helical tube bundle: *1*) theoretical relation for laminar flow in bundles of circular tubes with the same porosity; *2*) generalized relation [26] for a bundle of circular tubes with the same porosity; *3*) Blasius formula for a tube.

tube bundles is characterized by three specific regimes: laminar, laminar with macrovortices, and turbulent. The last also incorporates a transition regime, after which developed turbulent flow is produced with increasing Re. By analogy with spiral tubes, the onset of macrovortices in tubes with swirlers is characterized by the Dean number [56]:

$$De_{cr} = Re\sqrt{\frac{1}{0{,}5 + \frac{8}{\pi^2}\left(\frac{s}{d}\right)^2}} = 51 \tag{5.112}$$

If this equation is used for flow in helical tube bundles, then the onset of macrovortices will correspond to Re = 500 over a range of *s/d* = 8–12. From the experimental data in Fig. 5.9 it is seen that at Re = $(6\text{–}8)\cdot 10^3$ there exists the regime of fully developed turbulent flow. These values of Re are consistent with the Re that specifies the beginning of fully developed turbulent swirled flow in tubes, for which according to [56]

$$Re_{cr} = 38900\left(\frac{d}{s}\right)^{1,16} + 2300 \tag{5.113}$$

Thus, at Re = 10^3–$8 \cdot 10^3$ the laminar flow regime with microvortices was observed in bundles.

As seen in Fig. 5.5, the experimental results are in good agreement with the theoretical relations. At Re = $8 \cdot 10^3$ – $6 \cdot 10^4$, the maximum difference does not exceed 12% for tube bundles with s/d = 4.15. It should be noted that at $s/d < 6$ the slope of the experimental curves $\xi = f(\mathrm{Re})$ differs somewhat from the theoretical ones, which is attributed to the compound action of flow core losses that can be calculated only to some degree of approximation. The greater the flow swirling, the stronger is the influence of secondary flows. In this case, the relative contribution of these flows to the total losses increases with decreasing Re. Therefore, at $s/d < 6$, when the flow swirling increases, the increase of ξ, at moderate Re close to the transition regime, is somewhat higher than in the region of fully developed turbulent flow.

At Re > $6 \cdot 10^4$, the similarity flow regime occurs, therefore, within this range and the difference between experiment and theory increases.

Considering the good agreement between the experimental and predicted results at Re = $8 \cdot 10^3$–$6 \cdot 10^4$ and s/d = 4.15–12.45, and the small flow swirling effect on ξ (less than 20% at s/d > 12.45), Eq. (5.108) is recommended for calculating the hydraulic resistance coefficient in isothermal flow.

Within the range $10^3 < \mathrm{Re} < 6 \cdot 10^3$, the experimental data can be generalized, with a maximum deviation of ±6%, by the following equations:

for a bundle with s/d = 12.45

$$\xi = 6.76/\mathrm{Re}^{0.571} \tag{5.114}$$

for a bundle with s/d = 8.3

$$\xi = 8.51/\mathrm{Re}^{0.571} \tag{5.115}$$

for a bundle with s/d = 4.15

$$\xi = 5.37/\mathrm{Re}^{0.429} \tag{5.116}$$

As noted in Chapter 2, local hydraulic resistances were measured at three bundle sections.

For the fully developed turbulent flow region (at Re > 10^4), the local hydraulic resistance coefficients differ slightly (by less than 10%) for small twisting pitches for the 1st section, ξ being somewhat less for the 2nd and 3rd sections.

Coincidence of the data for the 2nd and 3rd sections indicates that hydrodynamic flow stabilization existed over the main section of the bundle.

The hydraulic resistance of bundles was also studied in the case of nonisothermal flow of a gas when heated over a temperature factor range, T_w/T_f = 1.1–1.4.

Figure 5.10 shows a comparison between the data and the isothermal curves. At Re > 10^4, for bundles with s/d = 12.45 and 8.3, the hydraulic resistance markedly decreases under heating conditions. This difference increases with increasing Re, amounting to the order of 10% at Re = $7 \cdot 10^4$, and this increase is attributed to the temperature factor effect (decreasing gas density and increasing gas viscosity near the wall), as is the case for a smooth channel.

For a bundle with a smaller twisting pitch (s/d = 4.15) a temperature factor effect on ξ was not observed. The decrease in the temperature factor impact on the hydraulic resistance at decreasing s/d is explained by the fact that flow nonisothermity exerts an influence mainly in the wall layer, whose relative contribution to the total hydraulic losses decreases sharply with increasing flow swirling at s/d < 8. A similar situation was observed in [26] where studies were made in tube bundles and in tubes with swirlers.

To calculate the hydraulic resistance coefficient ξ_H with gas heating within the limit T_w/T_f = 1–1.4, Re = 10^4–10^5, and s/d > 4.15, the following equation is recommended

$$\xi_H = \xi\psi \tag{5.117}$$

Here, ξ is determined by (5.108) for isothermal flow, and ψ is the empirical coefficient allowing for the temperature factor effect.

For s/d = 4.15–12.45

$$\varphi_x = \frac{u}{\sqrt{\tau_{xw}/\rho}}; \quad \bar{\varphi}_x = \frac{\bar{u}}{\sqrt{\tau_{xw}/\rho}} \tag{5.118}$$

where

$$\psi_0 = \left(\frac{T_W}{T_f}\right)^{-(\lg \mathrm{Re} - 4)} \tag{5.119}$$

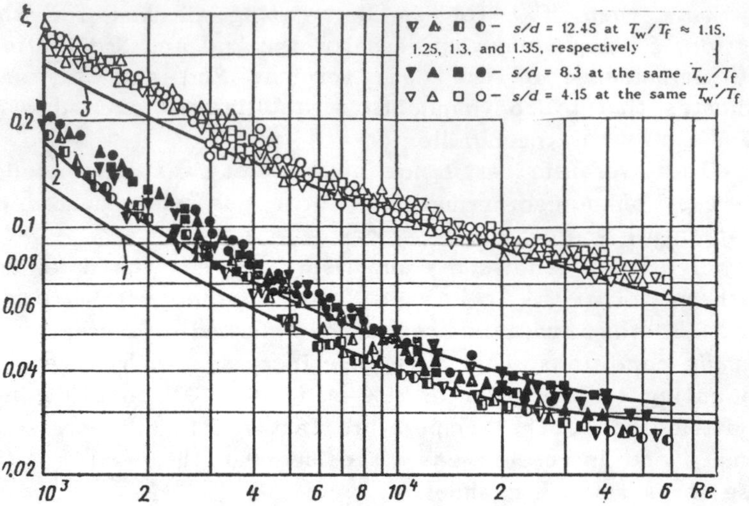

Figure 5.10 Friction factors with gas heating in helical tube bundles: *1, 2, 3*) isothermal relationships for s/d = 12.45, 8.3, and 4.15, respectively

For $s/d \leq 4.15$

$$\psi = 1 \tag{5.120}$$

for $s/d \leq 12.45$

$$\psi = \psi_0 \tag{5.121}$$

The decrease in the temperature factor effect with decreasing Re is due to the growing impact of free convection, which exerts a reverse influence on ξ. At Re < $6 \cdot 10^3$, the hydraulic resistance in the heated gas flow is higher than the isothermal case. Thus, at Re = 10^3–$2 \cdot 10^3$, the difference is 25%. In this case (vertical orientation of the experimental section, gas moves from top to bottom), free convection increases the hydraulic resistance, and this effect may be allowed for by the empirical formula:

$$\psi = 1 + 15\,\frac{(\mathrm{Gr}\cdot\mathrm{Pr})^{0.15}}{\mathrm{Re}^{0.8}} \tag{5.122}$$

where Gr = $g\beta d_{eq}{}^4 q/\nu^2 \cdot \lambda$ is determined using the heat flux density q and the equivalent diameter d_{eq}.

Results of hydraulic resistance coefficients for bundles with a large number of helical tubes. The hydraulic resistance of bundles with a large number of helical tubes (≥ 37) was examined using the generally accepted methods of experimental set-ups described in § 2.4. The air flowrate was measured by a washer calibrated on a gas holder. Pressure drops on the control sections were measured by liquid differential manometers and induction probes, and the pressure level was measured by standard capacitance-type sensors. The experiments covered the following parameter ranges: s/d = 6.5–35, Fr_M = 62–2440, $Re_f = 3 \cdot 10^3 – 5 \cdot 10^4$, T_w/T_f = 1.0–1.42, T_f = 237–467 K, $T_w \leq 621$ K, M = 0.03–0.27, N = 2.3. The maximum error associated with determining the hydraulic resistance coefficient was 9% according to the functional relation:

$$\xi = \xi\left(\mathrm{Re},\ \mathrm{Fr}_M,\ M,\ \frac{x_H}{d_{eq}},\ \frac{T_w}{T_f}\right) \tag{5.123}$$

The coefficient ξ was determined from the momentum equation for variable-density flow, considering that a helical tube bundle is a channel whose cross-sectional area is constant along the length, for which ρu_{mean} = const (x):

$$-dp = \xi \frac{\rho u_{mean}^2}{2} \frac{dx}{d_{eq}} + \rho u_{mean}\, du_{mean} \tag{5.124}$$

For a section of finite length l:

$$\Delta p = \frac{\rho_1 u_{mean\,1}}{2d_{eq}} \int_0^l \xi u_{mean}\, dx + \rho_1 u_{mean} \left(u_{mean2} - u_{mean1}\right) \tag{5.125}$$

For the fully developed flow section it may be assumed that ξ = const (x). Then, with $\int_0^l u_{mean} dx = (u_{mean1} + u_{mean2}) l_c/2$ and $\rho u_{mean} = G/F_f$, we obtain the expression:

$$\xi = \frac{\Delta p - \left(\frac{G}{F_f}\right)^2 \left[\left(\frac{1}{\rho_2}\right) - \left(\frac{1}{\rho_1}\right)\right]}{\frac{l_c}{d_{eq}} \left(\frac{G}{F_f}\right)^2 \frac{1}{2\rho_{mean}}} \tag{5.126}$$

where

$$\rho_{mean} = \frac{p_{mean}}{RT_{mean}}, \quad T_{mean} = \frac{T_1 + T_2}{2}, \quad p_{mean} = \frac{p_1 + p_2}{2}$$

Since in the experiments the air density and the flow velocity mostly deviated slightly from a linear relationship with length l_c, the coefficient ξ was determined by Eq. (5.126).

In processing the experimental data on the adiabatic airflow, gasdynamic tables and the method of successive approximations were used to determine the flow parameters along the length. From the calculated reduced velocity

$$\lambda = \frac{u}{a_{cr}} \tag{5.127}$$

where

$$a_{cr} = 18.3 \sqrt{T^*} \tag{5.128}$$

we have determined a thermodynamic flow temperature in a given bundle cross section, $T = \tau(T^*)$, air density $\rho = p/RT$, and velocity $u = G/\rho\ F_f$, where $\tau = \tau\ (\lambda)$. The stagnation temperature

$$T^* = T + \frac{u^2}{2c_p} \tag{5.129}$$

was measured in the receiver in front of the helical tube bundle where $u \approx 0$.

When the density and, hence, the flow velocity, deviated considerably from a linear relationship, the expression

$$\xi = \frac{d[p + \rho u_{mean}(u_{mean})]}{dx} \cdot \frac{2d_{eq}}{\rho u^2_{mean}} \tag{5.130}$$

was used to make the calculations in terms of local values of ρ and u, provided that the pressure distribution along the bundle length was known from experiment.

The results for ξ for bundles of oval- and three-blade-shaped tubes are presented in Fig. 5.11. Here, the

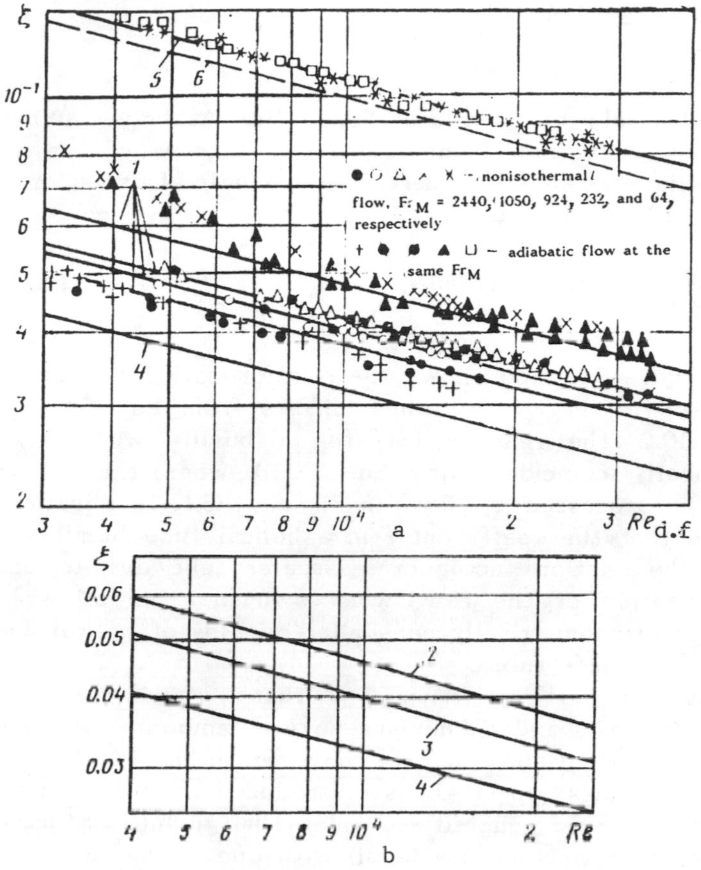

Figure 5.11 Friction factors versus Reynolds number determined in terms of oval hydraulic diameter (*a*) and three-blade (*b*) tube profiles: *1*) Eq. (5.131); *2,3*) experimental data for $Fr_M = 290$ and 1790 described by Eq. (5.131); *4*) Eq. (5.132); *5*) Eq. (5.134); *6*) relation from [51] for $Fr_M = 56$

plotted relations correlate well the data for adiabatic airflow in the stabilized flow length

$$\xi = B\xi_{tube} \tag{5.131}$$

where

$$\xi_{tube} = 0.3164\ Re_f^{-0.25}$$

$$B = 1 + 3.6\ Fr_M^{-0.357} \tag{5.132}$$

$$Re_f = \frac{\rho u_{mean} d_{eq}}{\mu}, \quad Fr_M = \frac{s^2}{d \cdot d_{eq}} \tag{5.133}$$

This relation is valid at $Fr_M \geq 100$. At $Fr_M < 100$, because of flow separation from the spiral surfaces of the tubes, a sharp increase in ξ is observed in a bundle. This region can be described by the equation

$$\xi = 0.3164\, Re_f^{-0.25} (1 + 3.1 \cdot 10^6 \cdot Fr_M^{-3.38}) \tag{5.134}$$

Figure 5.11 also shows the relation [51] for a helical tube bundle with $Fr_M = 56$, which differs from Eq. (5.134) by less than 10%. The relation [51] for a bundle with $Fr_M = 222$ practically coincides with Eq. (5.131) when the experimental data is processed by Eqs. (5.131) and (5.132). Equation (5.131) shows that the coefficient ξ in a helical tube bundle is larger than the friction factor in a circular tube due to additional turbulization of the flow when swirled. This allows one to generalize geometrically nonsimilar bundles of helical and oval- and three-blade tubes.

A universal resistance law that extends the modelling potentialities and diminishes the amount of necessary experimental studies may be obtained if some effective wall layer thickness δ [14] is introduced as a characteristic dimension. The quantity δ may be included formally by comparing resistance law (5.131) with one in the form:

$$\xi = \frac{0.266}{Re_\delta^{0.25}} \tag{5.135}$$

where the proportionality factor of 0.266 is equal to that in Blasius law (5.132) where the tube radius ($R = \delta$)

$$Re_\delta = \frac{\rho u_{mean} \delta}{\mu} \tag{5.136}$$

served as a characteristic dimension. Then, the equation

$$\frac{\delta}{d_{eq}} = \frac{0.5}{\left(1 + \frac{3.6}{Fr_M^{0.357}}\right)^4} \tag{5.137}$$

can be obtained to determine the quantity δ. Equation (5.137) is consistent with Eq. (3.30), obtained by measuring the velocity fields. At $Fr_M \rightarrow \infty$, the quantity δ tends to $d_{eq}/2$, and at $Fr_M \rightarrow 0$ the wall layer thickness $\delta \rightarrow 0$. The calculation results obtained from Eq. (5.131) also agree with the experimental ξ data for a nonisothermal flow in a helical tube bundle. No effect of M on ξ for an adiabatic compressible gas flow in helical tube bundles and in circular tubes over the investigated range of M was observed.

The effect of the hydrodynamic flow stabilization length in a helical tube bundle with axisymmetric flow from a large volume on ξ was investigated by calculating ξ using Eq. (5.126) for different control section lengths l_c at x_{ent}/d_{eq} = 3.75, 11.85, and 36.2. It appears that at $x_{ent}/d_{eq} \geq 3.75$ or $x_{ent}/2\delta \geq 12$ the mean value of ξ along the control section for all the analyzed versions is practically constant. Since different flow conditions past static pressure gauges in a helical tube bundle can cause the coefficient ξ to be underestimated by 5% at any flowrate [14], the quantity x_{ent}/d_{eq} = 3.75 should be considered as a minimum estimate of the hydrodynamic entrance length. The thermal entrance length of helical tube bundles is equal to $(l_{ent}/d_{eq})_{th} = 11$, unlike $(l_{ent}/d_{eq})_{th} = 50$ for a circular tube [33]. The hydrodynamic entrance length in a tube is $(l_{ent}/d_{eq})_{hyd}$ = 30 [33], i.e., it is approximately 70% less than the length (l_{ent}/d_{eq}). Thus, by analogy, it may be assumed that for helical tube bundles $(l_{ent}/d_{eq})_{hyd} \approx 8$. The relatively small values of the hydrodynamic and thermal entrance lengths in helical tube bundles is attributed to the levelling flow swirling action, which substantially expands the flow core along the entire bundle length and initiates a thin wall layer on the helical tubes.

The effect of the temperature factor on the coefficient ξ is shown in Fig. 5.12, from which it is seen that all the helical tube bundles examined over the range of T_w/T_f are governed by the relation:

$$\frac{\xi_{noniso}}{\xi_{adiab}} = \text{const} \left(\frac{T_w}{T_f}\right) \tag{5.138}$$

Only for the bundle with FR_M = 924 over a range of T_w/T_f = 1.2–1.42 was a deviation of the coefficient ξ observed within experimental accuracy. This leads to the conclusion that in helical tube bundles the variation of physical properties across

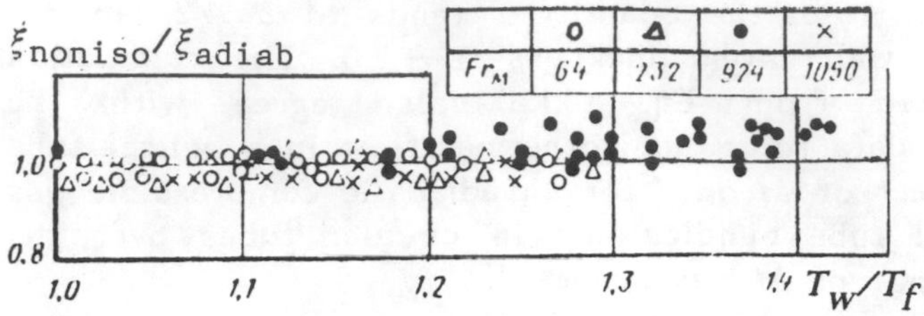

Figure 5.12 Friction factor ratio for a helical tube bundle versus temperature factor

the flow does not affect the coefficient ξ over the experimental parameter range examined. It should be noted that for tube flow, the variation of the physical properties over the thickness of the wall layer also exerts a substantially weaker influence on friction than on heat transfer.

The effect of the location of static pressure gauges on the coefficient ξ was examined on a helical oval tube bundle with $Fr_M \approx 330$. The roughness effect of these tubes began to manifest itself at Re ≈ $1.6 \cdot 10^4$ (Fig. 5.13). Two static pressure gauge locations were considered: at angles of a hexahedral shell and in the middle of the planes. In the first case, the hole diameter used to measure the static pressure was equal to $d_{s.p}/d_{eq} = 0.17$, and in the second case, $d_{s.p}/d_{eq} = 0.8$. In each cross section, the holes were assembled with a collector. In the first case, the number of holes was 6, and in the second, 4. For a bundle composed of 61 helical tubes, in the hexahedral tube shell with holes located at the hexahedron angles, the experimental ξ data agreed with Eq. (5.131) up to Re = $1.6 \cdot 10^4$, when there occurred a transition to the square resistance law (Fig. 5.13). When the static pressure holes with $d_{s.p}/d_{eq} = 0.8$ were located in the middle of the shell planes, the experimental data on ξ deviated markedly from Eq. (5.131) in the direction of a stronger dependence of ξ on Re (Fig. 5.13). The reason for this deviation is connected with the different flow conditions past the static pressure measuring holes. This difference is connected to both the effect of the vortex motion in the bundle cross section, thus resulting in a variability of the measured static pressure along the tube and the tube shell

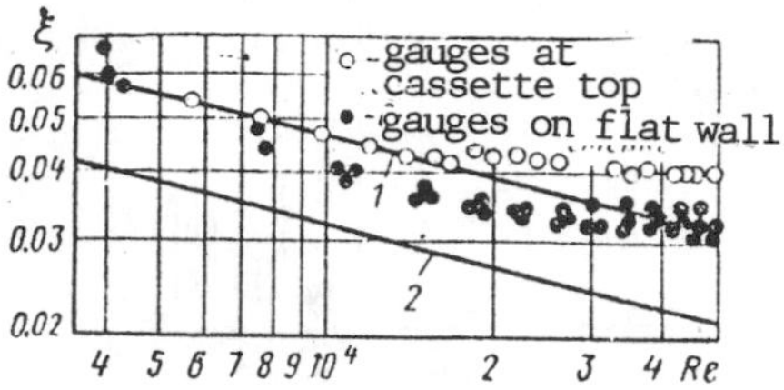

Figure 5.13 Effect of location of static pressure gauges on friction factor: *1*) Eq. (5.131) at Fr_M = 330; *2*) Blasius law.

perimeter [15], and to the fact that the hole diameter is comparable with the equivalent diameter at $d_{s.p}/d_{eq}$ = 0.8. As the experiments indicate (Fig. 5.13), preference is given to the location of the static pressure measuring holes at the angles of the hexahedral tube shell. This is because the velocity vector is practically parallel to the bundle axis, and the hole diameter $d_{s.p}$ is, almost by an order of magnitude, less than d_{eq} ($d_{s.p}/d_{eq}$ = 0.17).

Studying the nature of the transition from laminar to turbulent flow in helical tube bundles is of great importance not so much to obtain quantitative estimates of the coefficient ξ in these flow regions, as to understand the specific features of a swirling flow in complex-geometry channels. Such a study was made of a bundle composed of 7 helical tubes located in a hexahedral tube shell with s/d = 50 and Fr_M ≈ 13600 over a Reynolds number range of Re = 35–1350. The results are shown in Fig. 5.14, from which it is seen that laminar flow in this helical tube bundle exists up to $Re_{cr.1}$ = 90, at which point the flow begins to show some instability. The transition from laminar to turbulent flow is smooth and covers a considerable Reynolds number range (Re = 90–1000). Such a transition behavior in a helical tube bundle is similar to that in curvilinear channels when the loss of stability leads to another state of laminar flow associated with the occurance of secondary flows. However, because the flow structure in helical tube bundles is substantially more intricate than in a curvilinear channel, and because of the presence of additional flow turbulization sources, in a helical tube bundle the transition to a turbulent flow occurs at a smaller Reynolds number than in a straight circular tube [55]. The transition in a helical tube bundle differs from that in a curvilinear channel where the transition to turbulent flow takes place at larger

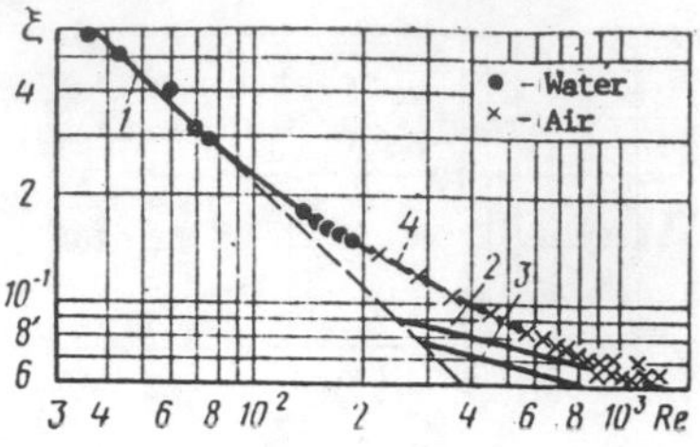

Figure 5.14 Transition from laminar to turbulent flow in a helical tube bundle with Fr_M = 13,600: *1*) Eq. (5.139); *2*) Eq. (5.131); *3*) Blasius law; *4*) transition line from laminar to turbulent flow regime

Reynolds numbers than in a straight circular tube. In this case, when the channel curvature increases, the transition to the turbulent flow regime shifts to larger Reynolds numbers.

The quantitative experimental data on ξ for the helical tube bundle in the laminar flow region may be described by the resistance law:

$$\xi_1 = \frac{A}{\text{Re}} \tag{5.139}$$

which differs from the Hagen-Poiseuille law for circular tubes (A = 64) because in this case A = 22.4. This difference is evidently connected with the fact that the constant A is a function of the channel shape when d_{eq} is used as the characteristic dimension, e.g., for triangular channels A = 48–54 [34]. It may be assumed that for helical different-shape tube bundles the value of A will be different.

5.4 EXPERIMENTAL HEAT TRANSFER RESULTS

Results of heat transfer for bundles composed of 19 helical tubes. Studies were made at Re = 10^3–$6 \cdot 10^4$, air pressure $(1.12–13) \cdot 10^5 N/m^2$, inlet flow temperature of ≈ 20°C, outlet flow temperatures 150–555°C, mean wall temperature T_w = 142–448°C, temperature factor T_w/T_f = 1.1–1.4, and heat flux density $q = 1.9 \cdot 10^3 – 1.9 \cdot 10^5$ W/m².

The mean heat transfer coefficient was determined at the section l_0 = 500 mm at distances x/d_{eq} = 48.6 and 35.6 from the entrance and exit, respectively. As seen from Fig. 5.15, the plot of Nu versus Re is smooth, which is attributed to the

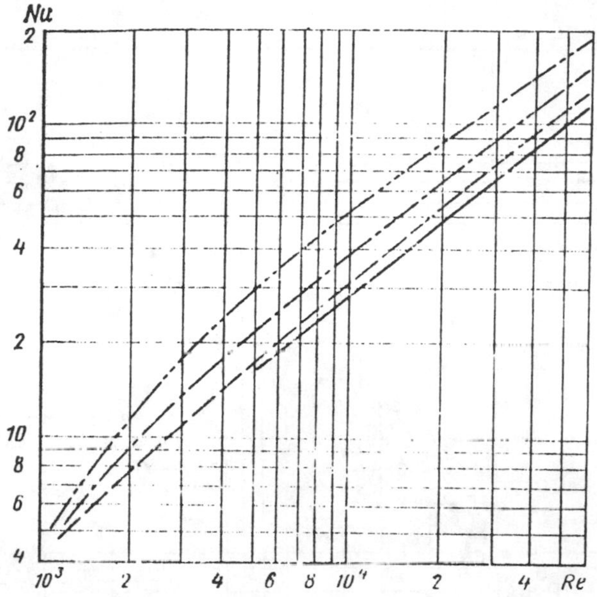

Figure 5.15 Mean Nusselt number for helical tube bundles: --- represents s/d = 12.43; —·—·— represents s/d = 8.3; —··—··— represents s/d = 4.15; ——— represents the formula for a tube

effect of secondary flows. As in the case of the friction factors, the experimental heat transfer results indicate a decrease in the temperature factor effect with increasing flow swirling degree (with decreasing s/d).

In the investigated range of T_w/T_f, the experimental data have the smallest scatter in processing $Nu_{reduced} = Nu\,(T_w/T_f)^n$ (Fig. 5.16), the exponent n being different for different values of s/d. For s/d = 12.45, n = 0.55, as in the case of heat transfer in smooth channels; for s/d = 8.3, n = 0.27; for a bundle with s/d = 4.15 a correction for the temperature factor effect was not required.

In the turbulent flow region (Re = $6 \cdot 10^3$–$6 \cdot 10^4$) the experimental data can be generalized by the relations:

$$Nu = 0.0232\, Re^{0.8} \left(\frac{T_w}{T_f}\right)^{-0.55} \qquad (5.140)$$

for s/d = 12.45;

$$Nu = 0.025\, Re^{0.8} \left(\frac{T_w}{T_f}\right)^{-0.27} \qquad (5.141)$$

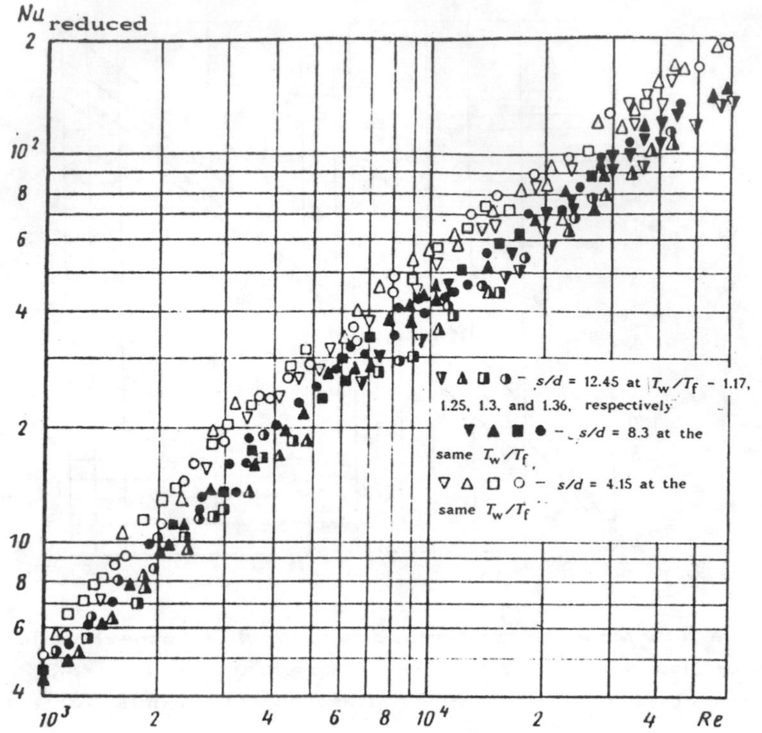

Figure 5.16 Average Nusselt number of helical tube bundles $Nu_{reduced} = Nu(T_w/T_f)^n$, where n = 0.063, s/d = 0.28

for s/d = 8.3;

$$Nu = 0.0851\,Re^{0.7} \tag{5.142}$$

for s/d = 4.15.

The change in the slope of Nu as a function of Re is attributed to the presence of macrovortices, whose relative contribution increases with growing flow swirling degree and with decreasing Re. A decrease in the temperature factor effect due to decreasing s/d is explained as follows. It is known that for heated gases the turbulent momentum transfer $(\overline{\rho u'v'})$ diminishes due to a decrease in the wall layer density. However, when the flow swirling degree increases, the turbulence generation in the flow core where the tangential velocity components undergo discontinuities is still of more significance for heat transfer. The greater the turbulence generation in the central channel zone where the temperature profile is flat, the smaller is the effect of wall layer nonisothermality.

The experimental data obtained using a correction for the temperature factor agrees fairly well with Eq. (5.110). The maximum difference between the experimental and predicted results does not exceed 8%. Considering that in the region with $s/d > 12.45$ the heat transfer changes only slightly with varying s/d, using Eq. (5.110) and the experimental results, the formula

$$Nu = 0.035\,Re^{0.75}\left[1+\frac{\pi^2}{0.5\frac{s}{d}\frac{s}{d_{eq}}}\right]^{0.4}\left[1+\frac{1.3}{\left(\frac{s}{d}\frac{s}{d_{eq}}\right)^{0.6}}\right]\cdot\left(\frac{T_w}{T_f}\right)^{-n} \qquad (5.143)$$

is recommended for calculating the heat transfer in helical tube bundles with $s/d > 4$ and $Re = 6\cdot10^3$–10^5, $T_w/T_f = 1$–1.14. Here, for $s/d = 4.15$–12.45

$$n = 0.55 - 0.0663\,(12.45 - s/d) \qquad (5.144)$$

for $s/d < 4.15$

$$n = 0 \qquad (5.145)$$

for $s/d > 12.45$

$$n = 0.55 \qquad (5.146)$$

In the present experiments, the values of the heat transfer in helical tube bundles with $s/d = 12.45$, 8.3, and 4.15 are 6, 14, and 42% higher than in circular tube bundles of the same porosity and are 29–73% higher than in a circular tube. This comparison is made at $T_w/T_f = 1$. Therefore, heat transfer improvement in helical tube bundles with $s/d < 12$, as compared to axial flow, increases with T_w/T_f. This occurs because the temperature factor in a swirled flow reduces the heat transfer less than in a smooth channel.

The mean heat transfer from 7 central tubes of a bundle was determined separately and differed slightly from the average for all 19 tubes of a bundle. The difference did not exceerd 7%; though the mean mass flow temperature in both cases was the same, the temperature was, indeed, somewhat higher at the bundle center.

For a low Reynolds number range of $Re = 10^3\text{–}4{\cdot}10^3$ the mean heat transfer data can be approximated by the relations:

$$Nu = 0.0048\ Re^{1.02} \tag{5.147}$$

for a bundle with $s/d = 12.45$,

$$Nu = 0.0021\ Re^{1.1} \tag{5.148}$$

for a bundle with $s/d = 8.3$,

$$Nu = 0.00089\ Re^{1.25} \tag{5.149}$$

for a bundle with $s/d = 4.15$.

The local heat transfer was determined in the cross section at distances $x/d_{eq} = 48.6$, 108, and 167 (cross sections *I, II, III*) from the entrance with respect to the parameters averaged for all 19 tubes of a bundle. In the turbulent flow region ($Re > 6{\cdot}10^3$), the local heat transfer decreases by 9–14% over the length from cross section *I* to cross section *II*. The local heat transfer in cross sections *II* and *III* is practically the same.

At $Re < 6{\cdot}10^3$, the local heat transfer variation as a function of distance from the entrance greatly increases, which is attributed to specific features of the flow dynamics in the experimental tube length. The distance from the start of the twisted surface of tubes to cross section *I* is 100 mm, i.e., $x/d_{eq} = 24$. This distance is probably insufficient to develop a fully swirled flow. In this case, the absolute velocity vector direction is still sufficiently close to the axial one. Therefore, when passing from protrusions to cavities on the helical tube surfaces, boundary layer separations are possible which enhance heat transfer and require less energy consumption compared to flow swirling. Hence, in cross section *I*, the local heat transfer coefficient is higher while the local friction factor is lower than in the remaining cross sections downstream. This specific feature occurs at large flow swirling degrees ($s/d < 10$) and relatively low Reynolds numbers (in the transition regime and in the laminar flow regime with macrovortices).

Based on experimental data on the local heat transfer and local friction factor, a length of $x/d_{eq} > 30$ may be assumed to be enough to establish thermal and hydrodynamic swirled flow stabilization in the turbulent flow regime ($Re > 6{\cdot}10^3$). At the stabilized length, the local heat transfer coefficients may be calculated by (5.143).

As mentioned above, one of the main advantages of a helical tube bundle is the presence of intensive mass transfer in the intertube space. In our experiments, the temperature nonuniformity over a bundle cross section of different tubes characterized the mixing efficiency of the heat carrier. At the same time, for bundles with s/d = 12.45 and 8.3, the temperature nonuniformity over a bundle cross section was the same. For a bundle with s/d = 4.15, this nonuniformity is 1.5-2 times as small (Fig. 5.17). This result shows agreement with the mechanism of changing hydraulic losses in the flow core $(\xi-\xi_w)$ (Fig. 5.6). As the calculation has shown, these losses sharply increase with increasing flow swirling degree for s/d = 8. Thus, when s/d decreases from 8 to 4.15, the flow core losses are increased twice.

The flow swirling effect on the levelling of temperature nonuniformities over a bundle cross section is also clearly seen from the scatter of the experimental data on different bundle cross sections (Fig. 5.17). As previously mentioned, in cross section *I* the flow is still slightly swirled; therefore, the temperature nonuniformity is higher here than in cross sections *II* and *III*, where flow swirling has achieved its maximum value. It is clear that this effect is most pronounced in the region of relatively low Reynolds numbers where turbulent pulsations have not yet developed and a larger x/d_{eq} compared to the one in the turbulent region is required to generate a swirled flow. Thus, proceeding from the theoretical and experimental data, it is considered that mixing is substantially improved by twisted tube bundles with $s/d < 8$.

Results of heat transfer for bundles with a large number of helical tubes. An experimental heat transfer study was made on bundles with a large number of helical tubes (≥ 37) with $Re_{df} = 2\cdot10^3-4\cdot10^4$, $Fr_M = 64-2440$, $T_w-T_f = 1-1.75$, $q_w = 0-64\cdot10^3$ W/m^2, $T_{f.in}$ = 287-304 K, $T_{f.out}$ = 287-467 K, and $T_w \le 621$ K. Different methods were adopted to process the experimental data. When an equivalent diameter was used as the characteristic dimension, the heat transfer experimental data (Fig. 5.18) were processed in the form:

$$Nu_{df}\left(\frac{T_w}{T_f}\right)^{0.55} = f\left(Re_{df}, \frac{x}{d}\right) \tag{5.150}$$

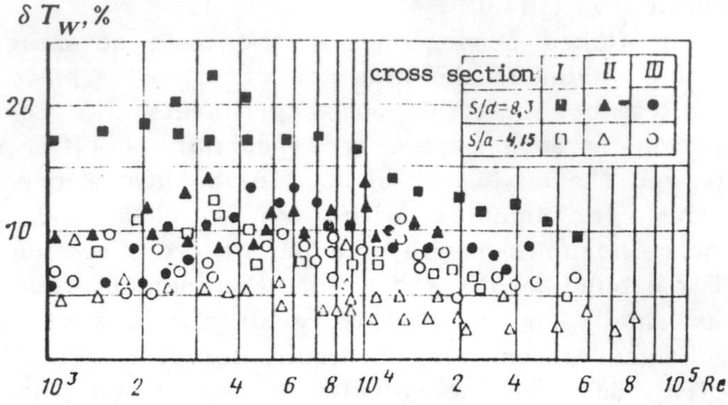

Figure 5.17 Temperature nonuniformity over the cross section of a helical tube

It is seen that Nu as a function of Re in the turbulent flow region for helical tube bundles is similar to the relation for circular tubes [33]:

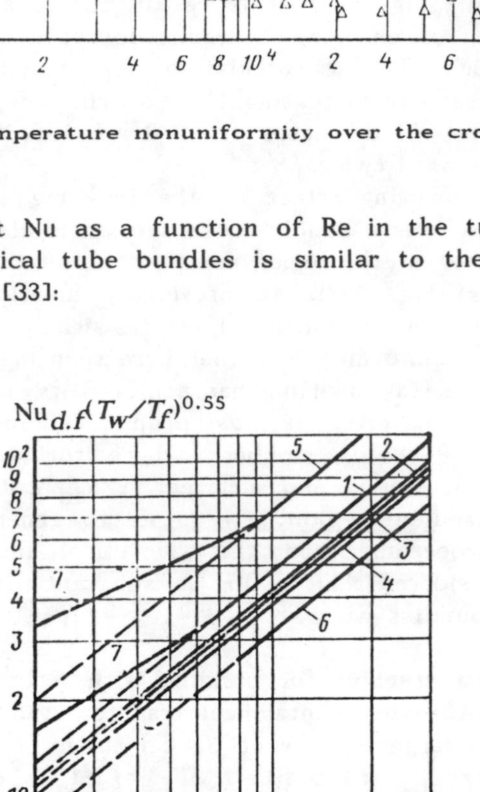

Figure 5.18 Nusselt number versus Re and Fr_M: *1–4*) experimental data for Fr_M = 232 and x/d_{eq} = 67, Fr_M = 924 and x/d_{eq} = 20, Fr_M = 1050 and x/d_{eq} = 83.6, Fr_M = 2440 and x/d_{eq} = 103, respectively, described y Eq. (5.151); *5*) the same for Fr_M = 64 and x/d_{eq} = 66.2 (Eq. (5.152)]; *6*) Eq. (5.151); *7*) experimental data described by Eq. (5.154)

$$Nu_{df} = 0.023\, Re_{df}^{0.8}\, Pr_f^{0.4} \left(\frac{T_w}{T_f}\right)^{-0.55} \tag{5.151}$$

and differs only by the constant muliplier that depends on Fr_M. The smaller is Fr_M, the larger is the Nusselt number, as compared to the value of Nu determined by (5.151). At Fr_M = 64 the fully developed heat transfer is described by the formula [13]:

$$Nu_{df} = 0.0521\, Re_{df}^{0.8}\, Pr_f^{0.4} \left(\frac{T_w}{T_f}\right)^{-0.55} \tag{5.152}$$

At Fr_M = 232–2440 the experimental heat transfer data for geometrically nonsimilar bundles can be generalized by a more general relation. Fr_M [14] was used as a characteristic number:

$$Nu_{df} = 0.023\, Re_{df}^{0.0}\, Pr_f^{0.4} \left[1 + \frac{3.6}{Fr_M^{0.357}}\right] \left(\frac{T_w}{T_f}\right)^{-0.55} \tag{5.153}$$

In this case it appears that Fr_M equally affects ξ and Nu. Since the experimental data for Fr_M = 64 do not obey Eqs. (5.131) and (5.153), starting from a certain value of $Fr_M < 232$, when Fr_M is decreased, the coefficient ξ for a helical tube bundle increases to a greater degree than the heat transfer coefficient. At the same time, this study points to the different effect of flow nonisothermity on the coefficient ξ and Nu. The effect of the temperature factor T_w/T_f on the heat transfer coefficient for a helical tube bundle is similar to that of T_w/T_f on the heat transfer coefficient for circular tubes [33]. Figure 5.18 illustrates a comparison between Eq. (5.153) and the experimental data on heat transfer only for some values of the longitudinal coordinate x/d_{eq} at the stabilized flow length. The bundle effect on heat transfer in a bundle with Fr_M = 924 at x/d_{eq} = 3.75–59.3 is shown in Fig. 5.19, where at x/d_{eq} = 20–59.3 the scatter of the experimental data is observed to be within ±15% with respect to the generalizing relations, which illustrates the different flow conditions past the thermocouples along the perimeter of a helical tube in a bundle.

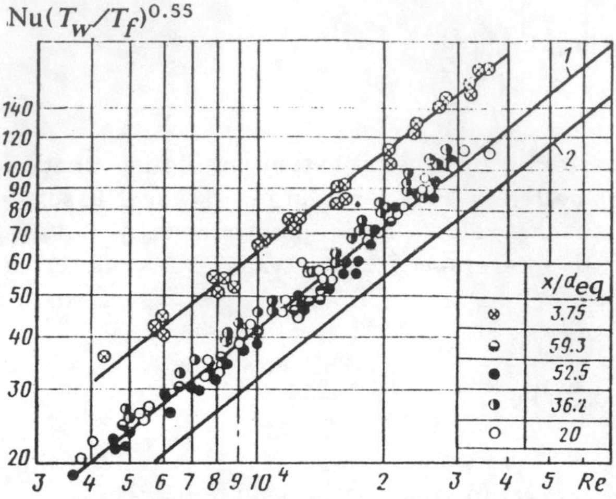

Figure 5.19 Experimental data on heat transfer in the intertube space of a heat exchanger with Fr_M = 924: *1*) Eq. (5.153); *2*) Eq. (5.151).

The experimental heat transfer data in the transition flow region may be generalized by the relation:

$$Nu_{d_f} = 83.5\, Fr_M^{-1.2}\, Re_{d_f}^{n}\, Pr_f^{0.4} \left(\frac{T_w}{T_f}\right)^{-0.55} \left(1 + \frac{3.6}{Fr_M^{0.357}}\right) \tag{5.154}$$

where at $Fr_M < 924$ the power

$$n = 0.212\ Fr_M^{0.194} \tag{5.155}$$

and at $Fr_M \geq 924$, $n = 0.8$. Equation (5.154) is indicative of substantially greater heat transfer enhancement in the transition flow region in a helical tube bundle than in the turbulent one (5.153).

The heat transfer law for helical tube bundles (5.153) may be reduced to a more universal form if some tube perimeter-mean wall layer thickness δ, (5.137), is introduced as a characteristic dimension. The quantity δ is, by definition, some integral geometrical characteristic of a bundle and is constant in the fully developed region. The experimental heat transfer data in helical tube bundles at x/d_{eq} = 20–103 was processed in the form:

$$Nu_{\delta m} = Nu\,(Re_{\delta m}, Pr_m) \tag{5.156}$$

where

$$Nu_{\delta m} = \frac{q_w\,\delta}{\lambda_m\,(T_w - T_{mean.f})} \tag{5.157}$$

$$Re_{\delta m} = \rho_m\,\delta\,\frac{1}{F_f}\int_F u\,dF/\mu_m \tag{5.158}$$

λ_m, ρ_m, and μ_m are the values of λ, ρ, and μ determined at a mean temperature in the boundary layer

$$T_m = \frac{T_w + T_{mean}}{2} \tag{5.159}$$

and are well described by the relation for circular tube flow when the tube radius is used as a characteristic dimension:

$$Nu_{\delta m} = 0.020\,Re_{\delta m}^{0.8}\,Pr_m^{0.4} \tag{5.160}$$

Agreement between the experimental data on heat transfer of helical tube bundles in the turbulent fully developed flow region and Eq. (5.160) is good, which indicates that this relation can be used to calculate the heat transfer in such bundles and that it is universal for cases when the physical properties approximately depend linearly on temperature. This condition is approximately satisfied with the physical properties of air and other diatomic gases.

For the transition flow region (at $Re_{\delta m} < 500$) the experimental data on heat transfer in a helical tube bundle can be generalized by the relation:

$$Nu_{\delta m} = 6.47\,Fr_M^{-0.845}\,Re_{\delta m}^{n}\,Pr^{0.4} \tag{5.161}$$

when the power n is determined by formula (5.155) at Fr_M < 924, and at $Fr_M \geq 924$ $n = 0.8$.

As previously mentioned, the friction factor (5.135) and the Nusselt number (5.160) laws may be reduced to a more convenient form $\alpha = \alpha(z)$, $\alpha_m = \alpha_m(z, z_m, \mathrm{Pr}_m)$ that extends the modelling potentialities and allows calculations by local flow characteristics. The dimensionless friction factor α and heat transfer coefficient α_m can be determined as follows [14]:

$$a = \frac{\tau_0}{\rho \bar{u}^2} \tag{5.162}$$

$$a_m = \frac{q_w}{\rho \bar{u} c_p (T - \bar{T})} \tag{5.163}$$

where the velocity $\bar{u}$ and temperature $\bar{T}$ at the external boundary of the wall layer are taken as characteristic velocities and temperatures, and the quantities

$$z = \frac{\mathrm{Re}_\delta}{a} = \frac{\bar{u}}{a\mu} \int_0^\delta \rho \frac{u}{\bar{u}} \left(1 - \frac{u}{\bar{u}}\right) dy$$

$$z_m = \frac{\mathrm{Re}_\theta}{a_m} = \frac{\bar{u}}{\mu a_m} \int_0^{\delta^{\cdot}} \rho \frac{u}{\bar{u}} \frac{T - \bar{T}}{T_w - \bar{T}} \, dy$$

are specially built Reynolds numbers [23]. Such a representation of the resistance and heat transfer laws allows [23] a revision of the hydrodynamic heat transfer theory (§ 5.1) by introducing the numbers z and z_m that take into account the difference in the thickness of the thermal and velocity boundary layers.

Figure 5.20 shows a comparison of the experimental data on the hydraulic resistance coefficient and Eq. (5.87). The coincidence is good, which indicates that this law may be employed to calculate the resistance coefficient of helical tube bundles. The experimental data on the heat transfer coefficient of helical tube bundles (Fig. 5.21) are also consistent with Eq. (5.94). This indicates that Eq. (5.94) may be used for calculating the heat transfer coefficient of such bundles.

5.5 THE SPECIFIC FEATURES OF HEAT TRANSFER STABILIZATION ALONG A BUNDLE

The heat transfer laws (§ 5.4) govern only locally averaged

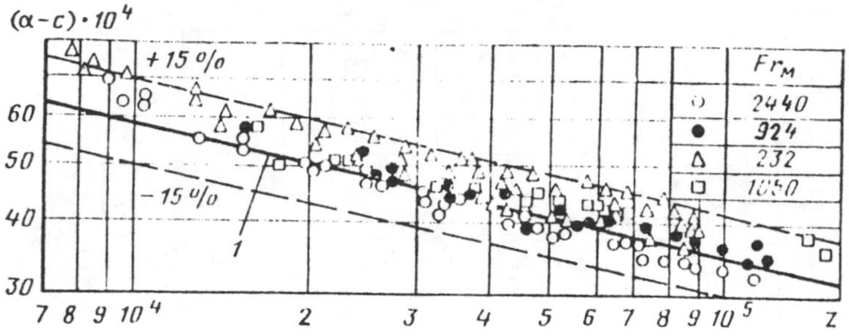

Figure 5.20 Correlation for the friction factor: *1*) Eq. (5.87)

heat transfer because an approximately ±15% scatter of the experimental data is observed along a bundle in the fully developed flow region. Variation of the heat transfer coefficient along the helical tube bundle is clearly seen from Fig. 5.19, where the experimental heat transfer results in a bundle with Fr_M = 924 are presented in the form of the functional equation (5.150). It is seen that the experimental heat transfer data for different x/d_{eq} are located equidistant, i.e., the Reynolds number practically does not influence the relationship between the Nusselt number and the quantity x/d_{eq} for the turbulent and transition flow regions. As mentioned above, the observed ±15% scatter of the experimental heat transfer data for different x/d_{eq} versus $\overline{Nu}$ may be attributed to different flow conditions of the heat carrier at different helical tube lengths, along which the thermocouples are embedded depending on relative location of helical tubes in a bundle. For the considered location of the tubes in a bundle, the locations of the embedding thermocouples on a tube wall in the oval might coincide with a region of through channels of a bundle, with contact points of adjacent tubes, or with intermediate regions of flow past tubes. Since the values of the local flow velocity past the tubes in the above regions are different and a mean mass velocity in the bundle cross section is taken as a characteristic velocity, it may be expected using the data processing procedures that Nu periodically changes along the bundle length with respect to $\overline{Nu}$ determined by Eqs. (5.153) or (5.160). At a distance x/d_{eq} = 20 from the bundle inlet when Fr_M = 924 (Fig. 5.19), the heat transfer coefficient is smaller than at a distance x/d_{eq} = 36.2. Also, at distances x/d_{eq} = 52.5 and 59.3 the coefficient has a smaller value than at x/d_{eq} = 20. If the origin of the longitudinal coordinate l is located in the outlet cross section of a bundle and the

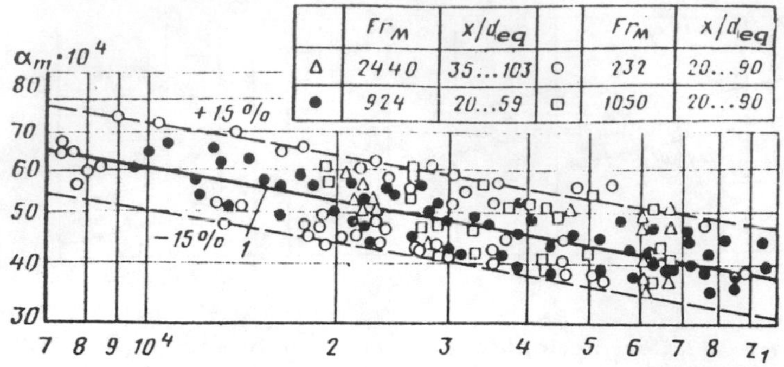

Figure 5.21 Correlation for the dimensionless heat transfer coefficient: *1*) Eq. (5.94)

direction upstream is taken as the positive direction of the coordinate l, then the relative coordinate l/s may be used to generalize the experimental data on local heat transfer from helical tube bundles with different Fr_M according to the following function (Fig. 5.22):

$$\frac{Nu}{\overline{Nu}} = 1 + 0.15\cos\frac{2\pi l}{s} \tag{5.164}$$

The deviation of the experimental points (Fig. 5.22) from the curve, Eq. (5.164), may possibly be attributed to a difference in the tube twisting pitch and bundle porosity with respect to the heat carrier. The closer the real bundle porosity is to a close-packed tube bundle, the closer is the agreement between the experimental data and Eq. (5.164). The functional equation (5.164) is valid only for an ordered tube packing, although a ±15% scatter of the heat transfer coefficient data practically does not depend on tube bundle arrangement. Then, the local heat transfer with fully developed tubulent flow in helical tube bundles may be described by the formula:

$$Nu = 0.023\left(1 + 0.15\cos\frac{2\pi l}{s}\right)Re^{0.8}\,Pr^{0.4}\left(1 + 3.6Fr_M^{-0.357}\right) \times \left(\frac{T_w}{T_f}\right)^{-0.55} \tag{5.165}$$

or

$$Nu_{\delta m} = 0.020\left(1 + 0.15\cos\frac{2\pi l}{s}\right)Re_{\delta m}^{0.8}\,Pr_m^{0.4} \tag{5.166}$$

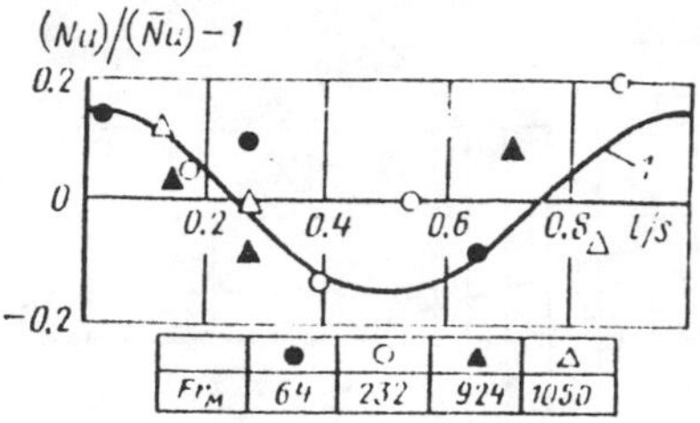

Figure 5.22 Effect of relative position of helical tubes on Nusselt number: *1*) Eq. (5.164)

The experimental data in Fig. 5.19 also point to an increased heat transfer coefficient in the entrance length of a helical tube bundle, as compared to the one in the stabilized flow length. When the data are processed in the form $Nu_{\delta m} = Nu(Re_{\delta m}, Pr_m, x/d_{eq})$, the effect of the entrance length on the heat transfer coefficient may be taken into account by introducing into Eq. (5.165) the expression:

$$C_m = \frac{\overline{Nu}_{\delta m}}{Re_{\delta m}^{0.8} Pr_m^{0.4}} = f\left(\frac{x}{d_{eq}}\right) \tag{5.167}$$

instead of a constant multiplier. Then, the experimental data for helical tube bundles with different Fr_M (Fig. 5.23) may be generalized by the power function:

$$C_m = 0.0426\left(\frac{x}{d_{eq}}\right)^{-0.287} \tag{5.168}$$

In this case, the relations for heat transfer with x/d_{eq} = 3.75–14 will be of the form:

$$Nu_{\delta m} = 0.0426\left(\frac{x}{d_{eq}}\right)^{-0.287} Re_{\delta m}^{0.8} Pr_m^{0.4} \tag{5.169}$$

$$Nu_{\delta m} = 0.0426\left(\frac{x}{d_{eq}}\right)^{-0.287}\left(1 + 0.15\cos\frac{2\pi l}{s}\right) Re_{\delta m}^{0.8} Pr_m^{0.4} \tag{5.170}$$

When processing the experimental data in the form:

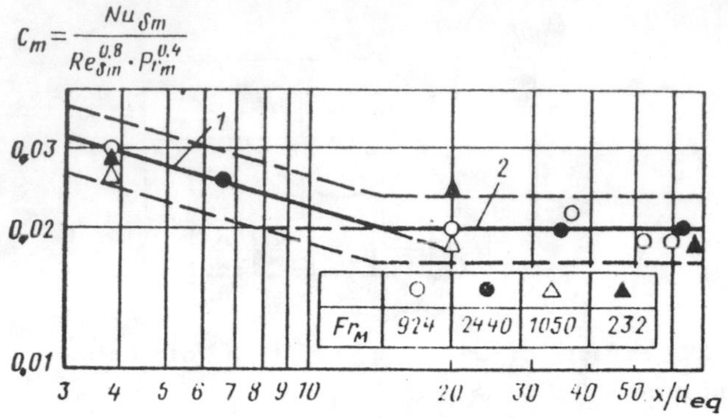

Figure 5.23 Effect of the entrance length on the local heat transfer coefficient: *1)* Eq. (5.168); *2)* Eq. (5.160)

$$\mathrm{Nu} = \mathrm{Nu}\left(\mathrm{Re}, \mathrm{Fr}_M, \frac{x}{d_{eq}}, \frac{T_W}{T_f}\right) \tag{5.171}$$

in Eqs. (5.153) and (5.165) the constant multiplier may be replaced by the expression:

$$C = 0.0490\left(\frac{x}{d_{eq}}\right)^{-0.287} \tag{5.172}$$

to calculate local heat transfer in the entrance length (x/d_{eq} = 3.75–14).

In the transition stabilized flow region at $\mathrm{Re}_{\delta m} < 500$, or at the number

$$\mathrm{Re} < 1000\,(1 + 3.6\,\mathrm{Fr}_M^{-0.357})^4\,\frac{T_w + T_{mean.f}}{2T_{mean.f}} \tag{5.173}$$

the heat transfer coefficient for a helical tube bundle is determined by (5.154) and (5.161), which, with regard to the specified features of heat transfer in the entrance length, will assume the form:

$$\mathrm{Nu} = 178\left(\frac{x}{d_{eq}}\right)^{-0.287}\left(1 + 0.15\cos\frac{2\pi l}{s}\right)\mathrm{Fr}_M^{-1.2}\,\mathrm{Re}^n\,\mathrm{Pr}^{0.4}\left(\frac{T_W}{T_f}\right)^{-0.55} \times (1 + 3.6\,\mathrm{Fr}_M^{-0.357}) \tag{5.174}$$

$$\mathrm{Nu}_{\delta m} = 13{,}8\left(\frac{x}{d}\right)^{-0.287}\left(1+0.15\cos\frac{2\pi l}{s}\right)\mathrm{Fr}_{M}^{-0.845}\,\mathrm{Re}_{\delta m}^{n}\,\mathrm{Pr}^{0.4} \tag{5.175}$$

and will differ by the multiplier $1 + 0.15 \cos (2\pi l/s)$ at the stabilized flow length when $x/d_{eq} > 14$.

The results obtained indicate that in the entrance bundle length, in the turbulent and transient flow regions, the temperature factor effect, T_w/T_f, on heat transfer is the same. It is similar to that in fully developed flow and is not affected by Fr_M. It is also revealed that the entrance length of helical tube bundles with axisymmetric flow entry is $x/d_{eq} = 14$ and does not depend on Fr_M.

The data obtained may be used to calculate the local heat transfer at the entrance and fully developed regions in longitudinal-flow helical tube heat exchangers having axisymmetric flow and ordered tube packing.

5.6 HEAT TRANSFER AND HYDRAULIC RESISTANCE IN A TWISTED BUNDLE OF HELICAL TUBES

A twisted bundle of helical tubes (Fig. 1.2) in longitudinal flow differs from the typical helical tube bundle (Fig. 1.1) not only by the presence of flow swirling in the bundle but also by tube packing. Typically, helical tubes are packed in a regular triangular array with the side equal to the maximum size (d) of the oval. In a twisted bundle, the triangular array, where close-packing of helical tubes is provided, is equilateral because the distance between the tube axes of adjacent rows is equal to $2d/\sqrt{3}$ and the distance between the tube axes of each rod is equal to d. Therefore, the porosity of such bundles (m_{twist}) with respect to the fluid is larger than that of typical helical tube bundles ($m_{straight}$), i.e., $m_{twist} > m_{straight}$. This influences the flow structure, fluid mixing, and other characteristics of flow and heat transfer. However, the twisting pitch of the tube rows relative to the bundle axis s_{twist} is the main parameter that determines the specific features of the behavior of these characteristics. In this case, the axial and rotational flow velocities have a maximum along the periphery with maximum helical tube twisting. Only the oval profile of the central helical tube of a bundle is twisted relative to the tube axis and this tube is streamlined with respect to the flow having minimum swirling and axial velocity.

Experimental studies of heat transfer and hydraulic resistance were made of a twisted bundle of helical tubes 1.5 m long with a constant twisting pitch s_{twist} = 650 mm. The helical tubes with d = 12.2 mm and δ = 0.5 mm had an oval profile twisted relative to the tube axis with a pitch s = 171 mm. There were 59 tubes in the bundle. The location of the tubes in the bundle over the concentric circles (Fig. 1.2) was as follows: 1, 6, 12, 17, 23 tubes in a row and a relative twisting pitch of the tubes in each row $s_{twist}/D = \infty$, 12.3, 11.6, 8.1, and 6.2, respectively, where D is the circle diameter of a given row. The mean bundle porosity was m_{twist} = 0.62.

The local modelling method was adopted to study the heat transfer. In this case, electric power was supplied to individual tubes of each row, and the heated length was 0.4 m. Thus, the heat transfer was investigated under hydrodynamic fully developed conditions, since the nonheated length l was equal to 90 d. The data processing procedure was essentially identical to that described in Chapter 7. The local heat transfer coefficient $\alpha = q_x/(T_{wx}-T_f)$, $Re_f = Gd_{eq}/(F\mu_f)$ and $Nu_{twist} = \alpha\, d_{eq}/\lambda_f$ were found from experiment. The incoming flow temperature T_f was taken as the characteristic temperature and $d_{eq} = 4F/\Pi$ = 14.19 mm as the characteristic dimension. Here, F is the flow cross section of a bundle. Studies were made within the range of Re = $1.4 \cdot 10^3 - 7 \cdot 10^4$. The experimental data were generalized in the form:

$$\frac{Nu_{twist}}{Nu_{straight}} = f\left(Re_f \cdot \frac{s_{twist}}{D}\right) \tag{5.176}$$

where $Nu_{straight}$ is the Nusselt number for the central helical tube of a bundle and is determined by the equations for a typical bundle. Equation (5.176) is shown in Fig. 5.24 as a plot of $Nu_{twist}/Nu_{straight}$ $f(s_{twist}/D)$ at different Reynolds numbers. It is seen that the twisting of helical tubes in a bundle enhances the heat transfer, and the larger the heat transfer the smaller is the relative twisting pitch of the tube rows of a bundle, s_{twist}/D. In this case, the smaller the Reynolds number, the larger is the heat transfer coefficient (Fig. 5.24).

Generally adopted methods [5] were used to investigate the hydraulic resistance coefficient of a twisted bundle of

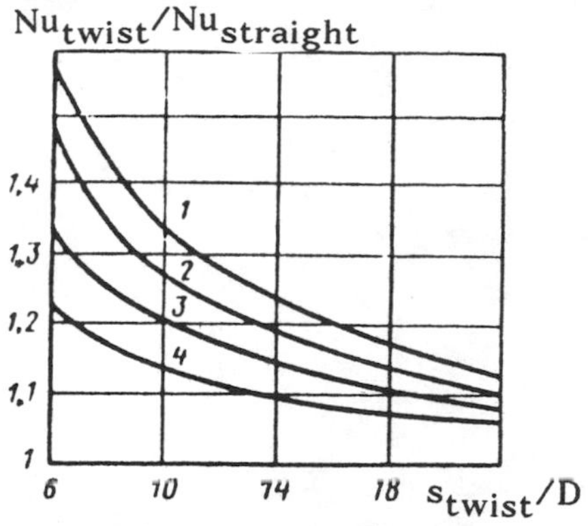

Figure 5.24 Nusselt number versus relative bundle twisting pitch and Reynolds number: *1–4*) experimental data for Re_f = $1.4 \cdot 10^3$, $5 \cdot 10^3$, 10^4, and $5 \cdot 10^4$, respectively

helical tubes. The results are presented in Fig. 5.25. It is seen that at Re = $1.4 \cdot 10^4$ the relation $\xi = \xi(Re\ )$ has an inflection, i.e., the variation of ξ versus Re_f is similar to ξ for a typical helical tube bundle. However, the value of ξ for a twisted bundle is somewhat higher than that for a typical bundle of helical tubes. The relations developed for heat transfer and hydraulic resistance may be employed to evaluate the efficiency of a heat exchanger with a twisted bundle of helical tubes.

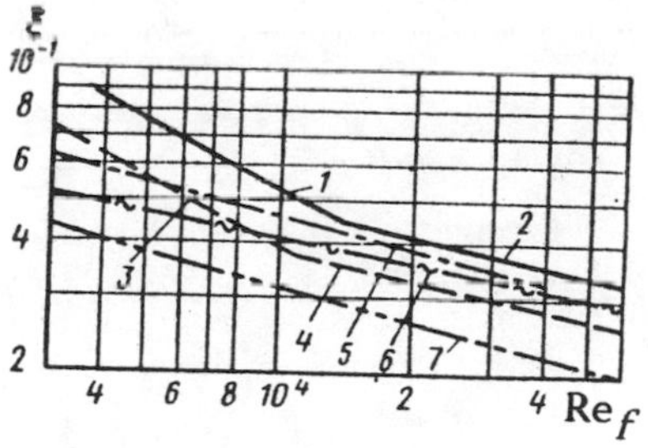

Figure 5.25 Friction factor versus Reynolds number for twisted bundles of helical tubes [5] with s_{twist} = 600 mm, m_{twist} = 0.63, and γ = 20°; m_{twist} = 0.62: *1–2*) experimental data described by $\xi = Re^n$ at C = 5.4, n = −0.5 and C = 0.37, n = −0.22, respectively; *3,4*) nontwisted bundle of helical tubes with s/d = 14 at C = 3.5, n = −0.5 and C = 0.29, n = −0.22, respectively; *5*) data of [14] for a bundle of helical tubes with s/d = 14; *6*) $\xi = 0.934 \cdot 10^{-2} + 0.316\ Re_f^{-0.25}$ for a twisted tube at D_{twist}/d = 9.45; *7*) smooth tube

CHAPTER

SIX

HEAT TRANSFER AND HYDRAULIC RESISTANCE OF A FLUID FLOWING INSIDE HELICAL TUBES AND IN CIRCULAR CHANNELS WITH AN INNER TWISTED TUBE

6.1 SPECIFIC FEATURES OF THE FLOW INSIDE SPIRALLY TWISTED TUBES

A force field imposed on the flow of a gas or liquid causes body forces whose value depends on the interaction between the force field and flow. Depending on the nature of the force field, gravitational, inertial, and electromagnetic fields of body forces are distinguished.

When a translational-rotational flow occurs in a circular tube, centrifugal forces appear. In this case, when the flow is steady, a moving gas or liquid particle is affected by: pressure forces that promote a motion; by viscous forces due to the relative motion of flow layers; and by body forces.

In a circular tube, both local and continuous swirling of a gas or liquid flow along the whole channel length may occur. Local flow swirling is created by swirlers at the tube inlet (these are blade, strip, screw swirlers) or by a tangential jet flowing into the tube through peripheral holes, etc. [57]. Continuous flow swirling along the entire tube is implemented by inserts (twisted strips, screws) and by changing the profile of the tube cross section.

A typical difference between tube flows with local and continuous flow swirling is that with local flow swirling affected by the viscous forces along the channel the flow structure continuously changes up to the moment when the rotational motion degenerates completely. The axial velocity profile recovery along the tube due to the decreasing flow swirling rate generates a radial velocity component and, thereby, a radial static pressure gradient. For continuous flow swirling, the dissipative forces do not decrease the degree of swirling.

In general, the swirled flows are spatial ones in the centrifugal force field. In such flows, the two and sometimes three velocity components (in the majority of cases, these are axial and tangential) are comparable and have transverse and longitudinal pressure gradients. These flows are characterized by pronounced turbulent pulsations. The investigation results and calculation methods for heat and mass transfer and hydrodynamic characteristics in axisymmetric channels with local flow swirling over a wide range of boundary and geometrical conditions are cited in [57].

With a steady-state rotational flow in a tube or curvilinear channel, liquid particles follow trajectories equidistant to the curvilinear channel walls. The centrifugal force acting upon a fluid element is balanced by the centripetal pressure gradient:

$$\frac{\rho u_\varphi^2}{r} = \frac{\partial p}{\partial r} \tag{6.1}$$

Random travel of a fluid element from its initial trajectory initiates excess mass forces that may disturb or stabilize the flow. From this point of view, it is possible to draw on the analogy between flows in the presence of rotation and flows in a temperature-stratified medium [52] and to show that the shear stress changes its sign with random travel of liquid particles over the radius to either side. This illustrates the opposing effect of the centrifugal mass forces, i.e., the excess mass forces have an active (at disturbance) and conservative (at stabilization) impact on the flow [52, 56]. In [52] it is emphasized that a circular liquid motion superimposed by a longitudinal motion produces an extra generation of kinetic pulsational energy in the longitudinal direction, and the rotation effect on the turbulence intensity in the radial and tangential directions (shear stress $\overline{u'_r \; u'_\varphi}$) is completely preserved in the presence of the longitudinal flow. The excess mass forces greatly affecting the flow may cause circulation flows in it, i.e., a rotational motion of individual flow elements. In this case, the conditions for the interaction between the flow and the curvilinear channel surface will also change.

Swirled tube flow follows a helical line; therefore, the wall region is characterized by a flow typical of that past a concave surface. It is known that near a concave surface the transport processes become more pronounced, and a region of active turbulizing impact of centrifugal forces appears. The turbulence generated by the mean motion increases and the quantity $\overline{u_r'^2}$ increases. Theoretical and experimental studies show that for flow around a concave surface, instabilities and macrovortices with axes parallel to the surface and with alternating rotation directions (Taylor-Görtler's vortices [55]) arise in a boundary layer. This is the reason for the heat and mass transfer enhancement, as compared to axial flows.

A flow motion over a helical line is also observed for flow inside a helical oval tube. The value of the excess mass forces in a helical tube depends on the geometry of the curvilinear channel (major-to-minor oval axis ratio, tube twisting pitch, i.e., its curvature) and on the velocity and temperature fields in the tube cross section. It is known that velocity profiles affected by body forces become more full and the temperature field is also distorted. The flow in a helical tube is similar to that in tubes with a strip or screw swirler where continuous flow swirling is implemented. The only difference is that it has no wall dividing the flow into two symmetric parts. The flow in a channel formed by a twisted strip and a tube wall may be considered as flow through a coil. Therefore, the developing field of body forces in the cross section of such a channel is able to create circulations consisting of a pair vortex covering the whole cross section. Such flows, with their quantitative characteristics of heat transfer and hydraulic resistance, similarity, and stability conditions are detailed in [56].

These features of the flow inside helical tubes must increase the hydraulic resistance and improve the heat transfer, as compared to a channel flow without centrifugal forces.

6.2 EXPERIMENTAL RESULTS ON HYDRAULIC RESISTANCE AND HEAT TRANSFER INVOLVING THE FLOW INSIDE HELICAL TUBES

The experimental results on heat transfer and friction were obtained on the apparatus shown in Fig. 6.1

A helical oval tube 1005 mm long with a maximum tube cross section d = 12.2 mm and with a wall 0.3 mm thick was

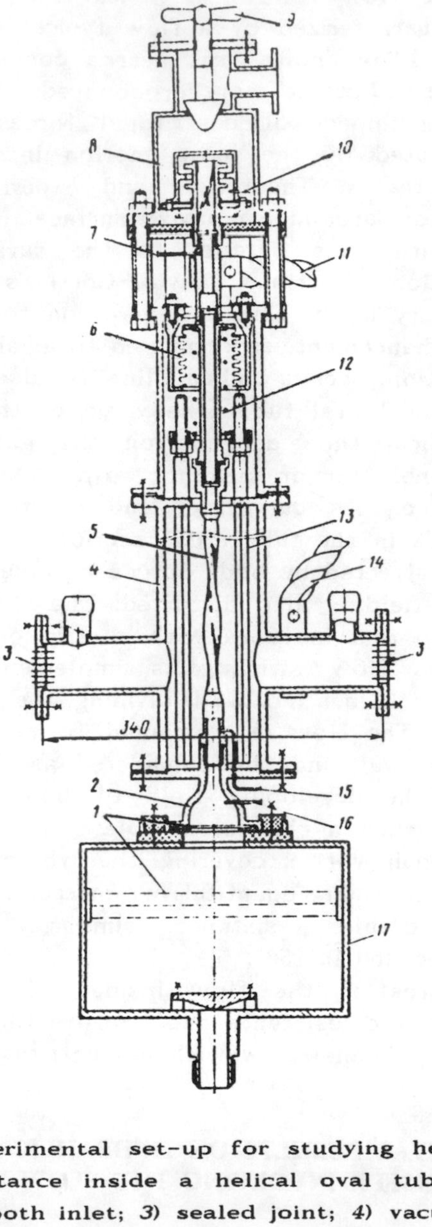

Figure 6.1 Experimental set-up for studying heat transfer and hydraulic resistance inside a helical oval tube: *1)* stabilizing screens; *2)* smooth inlet; *3)* sealed joint; *4)* vacuum chamber; *5)* calorimetric helical tube; *6)* siphon; *7)* upper pressure holes; *8)* mixer; *9)* flowrate controlling valve; *10)* outlet thermocouples; *11)* upper electric current supply; *12)* spring; *13)* reflecting screens; *14)* lower electric current supply; *15)* pressure holes; *16)* inlet thermocouples; *17)* upstream chamber

placed into a cylindrical vacuum chamber (*4*) to decrease the heat losses into the surroundings. In this chamber, the tube was encircled by three cylindrical coaxial polished sheet corrosion-proof 0.15 mm thick steel reflecting screens (*13*). A calorimetric helical tube was chosen so that the electric resistance variation along the 100 mm long section would not exceed 0.5%. The tube chosen was subjected to a preliminary 2 hr heating in air at 1090 K to stabilize the emissivity. The calorimetric tube was connected with the upstream chamber via a pipe connection. The inner pipe connection diameter was equal to that of a helical tube blank, i.e., the channel perimeters of the connecting pipe and the oval tube were identical. Air from the upstream chamber with compensating baskets was supplied to the tube through a smooth inlet. It then moved upwards and was discharged through a mixing device into the exhaust chamber and was then blown-out into atmosphere.

Holes of 0.35 mm diameter were drilled in the inlet and outlet connecting pipes to measure the tube pressure drop. The pressure in front of flow-measuring washers and at the experimental section inlet was measured by standard pressure gauges. Diaphragm and tube pressure drops were determined by means of differential cup-shaped pressure gauges filled with distilled water or by means of inductive pressure drop PD-1-type gauges. When heated, the calorimetric tube could freely move upwards.

The wall, screen, and air inlet and outlet temperatures were measured by chromel-alumel thermocouples. The tube inlet and outlet were equipped with three thermocouples in the connecting pipe and behind the mixing device. The wall temperatures of the calorimetric helical tube were measured on the outer surface over ten cross sections along the tube (in each cross section two thermocouples were embedded flush with the flat and rounded-off parts of the oval tube). Eight thermocouples were led out via sealed joints.

All electric signals of the thermocouples and the pressure drop gauge, as well as voltage drops on the tube and the shunt, were measured by an automatic measuring data acquisition system with a high-ohmic digital 0.01-grade voltmeter. All measurements were made at two voltages to eliminate any possible voltage gradient effect at the welded thermocouple heads.

The following experimental procedures were followed. Experiments were conducted in two runs. In the first run, the effect of Re on heat transfer was determined, i.e., a certain constant surface temperature of the calorimetric tube was maintained and the regimes were distinguished by the values of Re and the heat flux. In the second run, the impact of the variability of physical properties was specified, i.e., a certain constant airflow was maintained and the regimes differed in the surface temperature of the calorimetric tube and in the heat flux values. These experimental procedures were justified from the point of view of experimental accuracy and because that allowed the data to be obtained in a form most convenient for analysis and generalization. A further justification was that a number of researchers had established that the effect of variable physical properties of gas heat carriers on heat transfer in circular tubes under moderate heating conditions practically did not depend on Re.

In data processing, a mean mass flow temperature (bulk temperature) T_f was taken as the characteristic one. An equivalent diameter $d_{eq} = 4F/\Pi$, where the useful cross section F was determined by a hydrostatic weighing in calculating the tube wall thickness, served as the characteristic dimension. The thermal expansion of the calorimetric tube was taken into account in the heat flux calculations. In determining the linear variation of the convective heat flux component, account was taken both of heat losses through the screens and of longitudinal heat fluxes over the calorimetric tube wall.

Radiative heat transfer between the calorimetric tube and the screens was calculated by assuming they were coaxial and infinite cylinders.

The local mean mass flow temperature T_f was determined through the local enthalpy of the adiabatically frozen flow using a method of quadratic interpolation for enthalpy and a linear one for density. The mean hydraulic resistance coefficient was determined utilizing generally accepted methods (by the difference between the total pressure drop and the pressure drops on the connection pipes). The experimental data were processed on a computer, into which the automatically measured results were read using a punch card.

Experiments were made with two helical tubes having relative twisting pitches $s/d = 6.2$ and 12.2 over the parameter range of $\mathrm{Re} = 7{\cdot}10^3{-}2{\cdot}10^5$ and $T_w/T_f = 1.08{-}1.9$.

The hydraulic resistance of the oval tube was studied under isothermal flow conditions. The average hydraulic resistance coefficient over the length was determined. The experimental data in the Reynolds number range (Fig. 6.2) were generalized by:

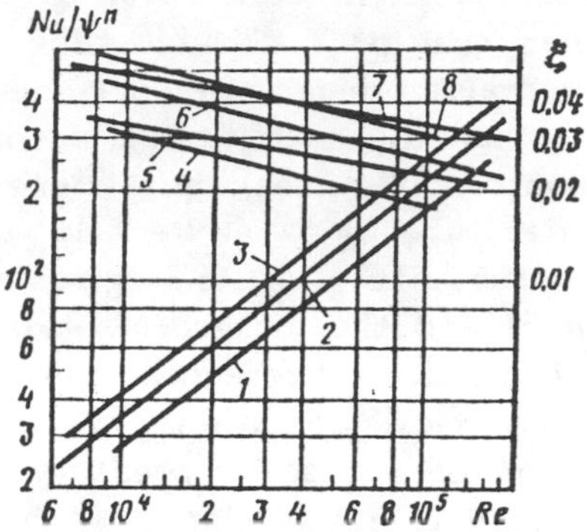

Figure 6.2 Experimental data on heat transfer and mean friction factor in helical tubes: *1, 4*) heat transfer and hydraulic resistance for circular tubes; *2, 5*) helical tube with s/d = 12.2; *3, 7*) for s/d = 6.2; *6, 8*) hydraulic resistance according to [24, 40]

for $s/d = 6.2$

$$\xi = 0.26 \ \mathrm{Re}^{-0.18} \tag{6.2}$$

for $s/d = 12.2$

$$\xi = 0.175 \ \mathrm{Re}^{-0.18} \tag{6.3}$$

This figure also shows the relation

$$\xi = 0.316 \ \mathrm{Re}^{-0.25} \tag{6.4}$$

for a circular tube, and the generalized relation [24, 40]

$$\xi = 0.316 \left[1 + 3.237 \left(\frac{s}{d} \right)^{-0.81} \right] \times \mathrm{Re}^{-0.25} \tag{6.5}$$

recommended for $s/d \geq 6.2$ and for the range of $\mathrm{Re} = 10^4$–10^5. In [24, 59], experiments were made with helical oval tubes having relative twisting pitches s/d = 16.69 and 6.213 and maximum cross sections d = 11.092 and 12.071, respectively. From Fig. 6.2 it is seen that the hydraulic resistance in a helical oval tube is higher compared to one in a circular tube and strongly depends on the relative tube twisting pitch. The hydraulic resistance coefficient at a twisting pitch s/d = 6.2 is 1.7 times higher than that of a circular tube and is within 5% of Eq. (6.5). The hydraulic resistance coefficient for s/d = 12.2 is considerably less than the coefficient ξ according to (6.2) for s/d = 6.2 (approximately 1.5 times), but is higher than the one for a straight circular tube by 5% at Re = 10^4 and by 24% at Re = 10^5. For this twisting pitch, Eq. (6.5) gives higher values, as compaerd to Eq. (6.3) and the disagreement is 35% at Re = 10^4 and 15% at Re = 10^5. Such differences may be attributed to different experimental flow conditions at the tube inlet and outlet, to the location of the static pressure holes, as well as to the different accuracies associated with determining the geometrical tube sizes. The noticeable disagreement between the obtained data on hydraulic resistance and those presented in [13] may also be caused by the accuracy in determining geometrical sizes. In [13], the experiments were made in a tube with s/d = 6.5 and a maximum tube cross section d = 6.33 mm, and the observed hydraulic resistance was 25–30% higher than for a circular tube, which is within the difference [24, 40] for a twisting pitch s/d = 12.2. However, in [13] the hydraulic resistance was investigated for nonisothermal flows. With air heating and T_w/T_f = 1.15–1.4, the hydraulic resistance decreased by 5–10%.

The flow inside an oval tube follows a helical line. It is, therefore, advisable to process experimental data on hydraulic resistance and heat transfer using the methods based on flow similarity in inertial force fields and using the relationships for coils and tubes with a strip swirler.

The centrifugal forces in turbulent flow substantially influence the flow behavior only with a considerable channel curvature. In this case, the flow will be turbulent with macrovortices. Depending on the shape of the channel cross section, secondary flows appear in the whole cross section or near the channel walls. Only for turbulent flow with macrovortices in a curvilinear channel does the hydraulic resistance increase and the heat transfer become enhanced [56].

The effect of mass forces on the flow development and transport characteristics is specified by the cross section shape. This can be taken into account by introducing the similarity numbers for a curvilinear channel as well as by introducing appropriate simplicities using an equivalent diameter as a characteristic dimension. Since an additional Reynolds number effect on heat transfer was not observed in tubes with a strip swirler, as compared to straight tubes, it may be assumed that this effect will also not occur in helical tubes.

According to [56], in the case of turbulent flow it is convenient to use the following two parameters to generalize the experimental data on heat transfer and hydraulic resistance for tubes with a strip swirler. These are Re and d/D, where Re is calculated in terms of a mean mass axial velocity and the equivalent diameter d is twice the distance between the flow points with maximum and minimum acceleration (for the complex-geometry channels it is proposed to use d_{eq}) and D is the curvature diameter of the axial channel line. Then, proceeding from geometrical considerations, when d is replaced by d_{eq}, we have the expression for D/d_{eq}:

$$\frac{D}{d_{eq}} = 0.5 + \frac{8}{\pi^2}\left(\frac{s}{2d_{eq}}\right)^2$$

for s/d = 6.2, D/d_{eq} = 18.6; and for s/d = 12.2, D/d_{eq} = 79.56.

The experimental results on hydraulic resistance for helical tubes were processed (as in the case of bent tubes and tubes with strip swirlers) in the form:

$$\xi\left(\frac{D}{d}\right)^m = f\left[\mathrm{Re}\left(\frac{d}{D}\right)^2\right]$$

Since the Reynolds numbers corresponding to the onset of turbulent flow with macrovortices in a helical tube with a twisting pitch s/d = 6.2 and 12.2 are approximately equal to $7\cdot10^3$ and $5\cdot10^3$, respectively, the experiments in helical tubes were made under conditions of turbulent flow with macrovortices.

The experimental data on hydraulic resistance inside a helical oval tube may be correlated as:

$$\xi\left(\frac{D}{d_{eq}}\right)^{0.67} = 0.71\left[\mathrm{Re}\left(\frac{d_{eq}}{D}\right)^2\right]^{-0.2} - 0.008 \qquad (6.6)$$

or

$$\xi = \frac{0.71}{\mathrm{Re}^{0.2}}\left(\frac{d_{eq}}{D}\right)^{0.27} - 0.008\left(\frac{d_{eq}}{D}\right)^{0.67} \qquad (6.7)$$

Figure 6.3 presents the experimental data, which has an accuracy of better than 11%. The nature of the relationship between the hydraulic resistance coefficient and the characteristic similarity numbers has also been observed for helically rolled tubes [41], spiral-corrugated tubes [28], and for tubes with strip swirlers [47]. An increase in the hydraulic resistance coefficient for steady flow is stipulated by the fact that the pressure losses in the oval tube are associated with friction and additional flow turbulization due to swirling, which is also typical of tubes with swirlers shaped as a twisted strip.

When the coefficient ξ for a helical tube is determined by a difference between the total pressure drop and the pressure drops on the connecting pipes, the inlet and outlet losses due to a varying channel shape have been estimated by the methods presented in [22], while the losses for forming and damping the tangential velocity component on the connecting pipes have been estimated by the semiempirical methods presented in [56]. These estimates have shown that the total contribution of the above losses in the Reynolds number range considered is from 10 to 16%.

In investigating heat transfer in a helical oval tube it is important to determine the exponent of the temperature factor T_w/T_f. The exponent of the temperature factor for a smooth tube may be used only as a first approximation since under the

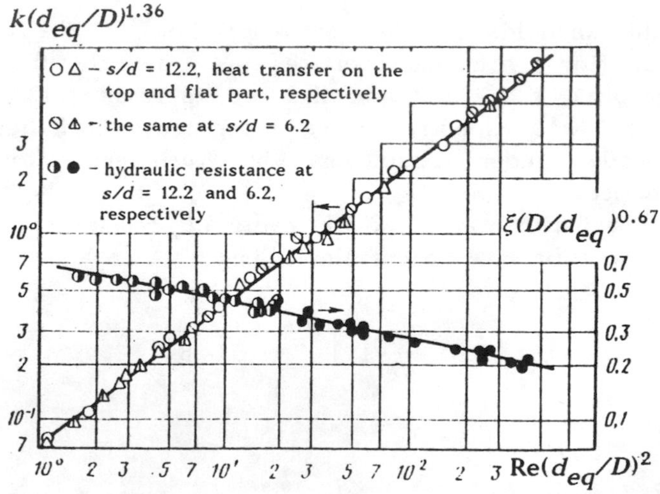

Figure 6.3 Generalized data on hydraulic resistance and heat transfer in helical tubes

action of the centrifugal forces in a helical tube additional turbulent generation occurs.

In the analyzed range of the parameter T_w/T_f = 1.08–1.9, the exponent n of the temperature factor T_w/T_f as a function of a relative length x/d_{eq} at a twisting pitch s/d = 12.2 is governed by the formula:

$$n = -\left(0.22 + 0.0023\frac{x}{d_{eq}}\right)\left[1 - \exp\left[-0.12\frac{x}{d_{eq}}\right)\right] \tag{6.8}$$

For the stabilized (fully developed) heat transfer length this relation reduces to:

$$n = -\left(0.22 + 0.0023\frac{x}{d_{eq}}\right) \tag{6.9}$$

For a twisting pitch s/d = 6.2, the exponent n of T_w/T_f increases along the length up to $x/d_{eq} \approx 30$ and then acquires a steady value equal to −0.17. The large scatter of experimental data up to $x/d_{eq} \approx 20$, and the behavior of n for a tube with s/d = 6.2 may be attributed to flow separations from the helical tube surface. Positive values of n are also obtained in a channel with a barrier for a separated flow region behind the barrier [10]. In the range considered of the temeprature factor $\psi = T_w/T_f$, the exponent n as a function of relative length at $x/d_{eq} > 30$ with relative twisting pitches s/d = 6.2–12.2 is described by:

$$n = -0.17 - 0.27\cdot 10^{-5}\left(\frac{x}{d_{eq}}\right)^{1.37}\left[\left(\frac{s}{d_{eq}}\right)^{2.1} - 109.6\right] \tag{6.10}$$

From the above it is seen that the effect of a temperature factor in helical tubs differs strongly from that in circular tubes. For stabilized turbulent flow at T_w/T_f = 1.5 for a helical tube with s/d = 6.2, using the exponent for the temperature factor typical of a circular tube underestimates the heat transfer results by 12%.

Local heat transfer data are shown in Figs. 6.2 and 6.4. Some difference is observed in heat transfer on the rounded-off part of the oval and on the flat surface. This difference can be attributed to the impact of the centrifugal forces due to flow swirling in a helical oval tube. The wall temperature on the flat part of the oval tube is several degrees (2–4°C) higher than the rounded-off surface at a mean temperature of ≈ 45°C. As seen in Fig. 6.4, the fully developed heat transfer in the Reynolds number range considered starts to form at a relative length $x/d_{eq} \geq 17$. At smaller relative lengths, i.e., $x/d_{eq} = 5.3$ and 8.2, the fully developed heat transfer begins at $Re = 3 \cdot 10^4$ and $2 \cdot 10^4$, respectively. The same situation is also observed at a twisting pitch $s/d = 6.2$.

In general, the hydrodynamic and thermal entrance lengths may be different and be affected both by turbulent flow characteristics and by the Prandtl number. For gas flows ($Pr \approx 1$), according to the data of various authors, it is known that the hydrodynamic and thermal entrance lengths are approximately the same.

In straight tubes the hydrodynamic entrance length in the turbulent regime is 20–40 diameters. The centrifugal force superimposed on the viscosity forces promote an axial velocity profile redistribution at a considerably smaller length, thus decreasing the hydrodynamic entrance length. In [56] it is shown that in a curvilinear rectangular channel the velocity profile transformation to a stable form is observed at a distance of 6.5 equivalent diameters from the bend start. Several reasons explain the decrease of the hydrodynamic entrance length in this case. One of them is associated with upstream disturbances due to the channel curvature [56]. In this case, the flow structure begins to form already in the straight part of the channel. Another reason is associated with the developing secondary flows, which generate the radial component of the pulsational velocity near the concave wall, thus promoting enhancement of the transfer processes.

The results for local heat transfer inside a helical oval tube are plotted as $Nu/\psi^n = f(Re)$ in Fig. 6.2. It is seen that the exponent of Re is equal to 0.8 both for a circular tube and for other devices that improve heat transfer, for example, for tubes with twisted strip inserts. One exception is spiral-corrugated tubes, where the exponent of Re somewhat exceeds this value [28].

For stabilized heat transfer, the data on the local heat transfer along a helical tube are well approximated by the expressions:

for $s/d = 12.2$

$$Nu = 0.025\, Re^{0.8}\, Pr^{0.4}\, \psi^{-(0.22+0.0023x/d_{eq})} \tag{6.11}$$

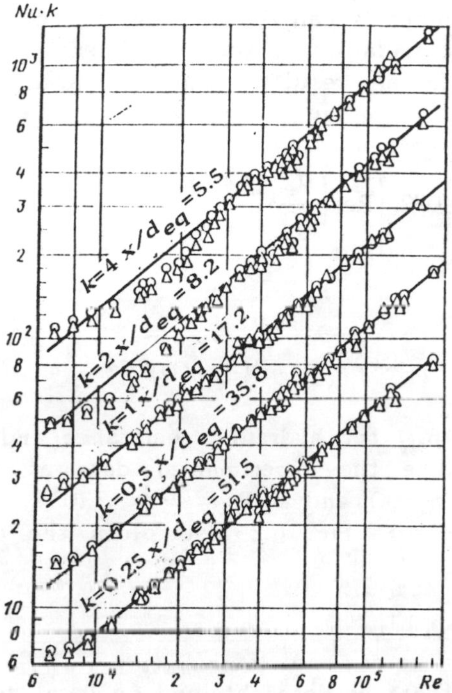

Figure 6.4 Heat transfer in a helical oval tube with s/d = 12.2 at different lengths x/d_{eq}. Circles represent the rounded-off part of the oval; triangles represent the flat surface

for s/d = 6.2

$$\mathrm{Nu} = 0.0294\,\mathrm{Re}^{0.8}\,\mathrm{Pr}^{0.4}\,\psi^{-0.17} \tag{6.12}$$

Figure 6.3 illustrates the heat transfer results for helical tubes. These data were generalized by an equation proposed in [56] for tubes with strip swirlers:

$$k\left(\frac{d}{D}\right)^m = f\left[\mathrm{Re}\left(\frac{d}{p}\right)^2\right]$$

where

$$k = \frac{\mathrm{Nu}}{\mathrm{Pr}^{0.4}(T_w/T_f)^n}$$

Equation (6.10) for the power n is used to take into account the airflow nonisothermity in the form of $(T_w/T_f)^n$. The Nusselt number is calculated using an equivalent diameter. A satisfactory correlation of the experimental data is obtained at m = 1.36. The air results (within an accuracy of 10%) are described by the formula:

$$k\left(\frac{d_{eq}}{D}\right)^{1.36} = 0.072\left[\mathrm{Re}\left(\frac{d_{eq}}{D}\right)^2\right]^{0.76} \tag{6.13}$$

or

$$\mathrm{Nu} = 0.072\,\mathrm{Re}^{0.76}\,\mathrm{Pr}^{0.4}\,(T_w/T_f)^n\,(d_{eq}/D)^{0.16} \tag{6.14}$$

Equations (6.7) for hydraulic resistance and (6.4) for heat transfer generalize the experimental data at s/d = 6.2–12.2, $\mathrm{Re}\cdot(d_{eq}/D)^2$ = 1.5–500, and at Re = $7\cdot10^3$–$2\cdot10^5$.

For comparison, Fig. 6.2 also plots the equation for a circular tube

$$\mathrm{Nu} = 0.021\,\mathrm{Re}^{0.8}\,\mathrm{Pr}^{0.4}\,\psi^{-0.55} \tag{6.15}$$

It is seen that the data obtained on local heat transfer for a helical oval tube lie above those for a circular tube and exceed them by as much as 1.2 times for s/d = 12.2 and 1.4 times for s/d = 6.2. In [40], the heat transfer over the parameter ranges s/d = 6.2–16.69, Re = $6\cdot10^3$–10^5, and T_w/T_f = 1–1.55 is generalized by the relation

$$\mathrm{Nu} = 0.019\left[1 + 0.547\left(\frac{s}{d}\right)^{-0.83}\right]\mathrm{Re}^{0.8} \tag{6.16}$$

The Nusselt numbers calculated by Eq. (6.16) for a twisting pitch s/d = 12.2 yield data underestimated by 9.6%, as compared to those calculated by Eq. (6.13). For s/d = 6.2 the underestimation of the calculated data is 20.3%. This divergence may be attributed to the fact that during data processing [40], the temperature factor was taken into account in a manner adopted for a circular tube. This is supported by the fact that at smaller relative twisting pitches a marked divergence between the heat transfer coefficient and the one determined in [40] is larger.

An attempt was made to allow for the impact of centrifugal forces on heat transfer in a helical oval tube in [13] utilizing Gr/Re^2. In [13] it is shown that Gr/Re^2, after being transformed, may be reduced to the parameter (T_w/T_f-1), which enables one to account for the centrifugal forces. In this case, the experiments made in the parameter range of Re = $2\cdot10^4$–$2\cdot10^5$, T_w/T_f = 1–1.6, and s/d = 6.5 were generalized by the relation:

$$Nu=0.0182\,Re^{0.8}\left(\frac{T_w}{T_f}\right)^{-0.55}\exp\left[1.95\left(\frac{T_w}{T_f}-1\right)^2\right] \qquad (6.17)$$

which differs from Eq. (6.15) by the multiplier

$$\exp\left[1.95\left(\frac{T_w}{T_f}-1\right)\right]$$

Good agreement between the calculations by Eqs. (6.13) and (6.17) is observed only at $T_w/T_f \approx 1.5$. At smaller values of T_w/T_f the divergence is considerable. Equation (6.17) is, by its structure, more suitable for a temperature-stratified media. The temperature ratio T_w-T_f does affect the transfer processes in a centrifugal force field; however, this influence is not decisive, as the action of centrifugal forces also occurs in isothermal flows.

6.3 HEAT TRANSFER IN A CIRCULAR CHANNEL WITH AN INNER HELICAL TUBE

The works of M. A. Nyyamira and Yu. V. Vilemas [7, 37] are devoted to an experimental study of the local heat transfer from a helical oval tube mounted at the center of a circular tube. Studies were made at five values of relative twisting pitch: s/d = 6.16,, 8.26, 11.8, 24.4, and 49.0 for almost a constant maximum diameter of the oval equal to $d \approx 12.2$ mm. The investigated helical tubes were placed into a circular tube with an inner diameter D = 28 mm. In this case, the equivalent diameter utilized as a characteristic dimension was almost constant and equal to $d_{eq} \approx 18.4$ mm. The schematic of the experimental set-up is detailed in [7]. Two thermocouples were embedded flush with the inner surface of the helical oval tube in five cross sections (one thermocouple is on the flat part, and another is on the rounded part) to measure surface temperatures and to determine voltage drops.

When generalizing the experimental data, the Nusselt number and the parameter T_w/T_f were averaged over the perimeter. It is noted that at the maximum Reynolds number the heat transfer on the rounded part of the oval was 13–18% higher by 4–6% as compared to the flat part while at the minimum Reynolds number. Experiments were conducted in the range of Re = $5 \cdot 10^3 – 4 \cdot 10^5$, and the temperature factor T_w/T_f was 3.

The experimental heat transfer data indicated that at the stabilized heat transfer length the heat transfer from helical tubes with a relative twisting pitch s/d = 24.4 and 49.0 is the same as from a rod, i.e., such tube twisting does not affect the perimeter-mean heat transfer. A different situation was observed for relative tube twisting pitches s/d = 6.16, 8.26, and 11.8. In the above Reynolds number range, heat transfer of the rod was divided into two zones: Re < $4 \cdot 10^4$ with different Re exponents. The exponent is affected by the twisting pitch of the tube. This influence is expressed well by the following relations:

at Re < $4 \cdot 10^4$

$$m = 0{,}8\left[1 - 0{,}35\exp\left(-0{,}105\frac{s}{d}\right)\right] \tag{6.18}$$

at Re > $4 \cdot 10^4$

$$m = 0{,}8\left[1 + 0{,}06\exp\left(-0{,}074\frac{s}{d}\right)\right] \tag{6.19}$$

The exponent m of the Re tends to 0.8 with increasing relative twisting pitch.

The variability of physical properties can be taken into consideration by using the parameter T_w/T_f. The dependence of the exponent n of the temperature factor on a relative length x/d_{eq} for a tube, circular channels, and in a circular channel with an inner helical tube is qualitatively the same.

For x/d_{eq} > 20, the experimental data on helical oval tubes can be described by the relation:

$$n = -\left[0.26\left(1 - 0.8\,e^{-0.08s/d}\right) + 0.0014\frac{x}{d_{eq}}\right] \tag{6.20}$$

The local heat transfer results can be approximated by the relation:

$$Nu = C\,Re^{m}\,Pr^{0.4}\left(\frac{T_w}{T_f}\right)^{n} \qquad (6.21)$$

It is emphasized that for a cylindrical rod fully developed heat transfer starts at x/d_{eq} = 30–35, and for a helical oval tube at 10–35. For stabialized heat transfer the following values of a constant C (Table 6.1) were obtained.

Table 6.1

s/d	6.16	8.26	11.8	24.4	49.0
$Re < 4\cdot 10^4$	0.124	0.0797	0.0474	0.027	0.0223
$Re > 4\cdot 10^4$	0.0188	0.0175	0.0166	0.0193	0.0221

Figure 6.5 illustrates the heat transfer change as a function of tube twisting pitch, as compared to the appropriate circular channel. It is also seen that the effect of the tube twisting pitch on heat transfer is complex in nature. Heat transfer improvement is observed in all cases at s/d < 10. The twisting efficiency at large temperature differences proves to

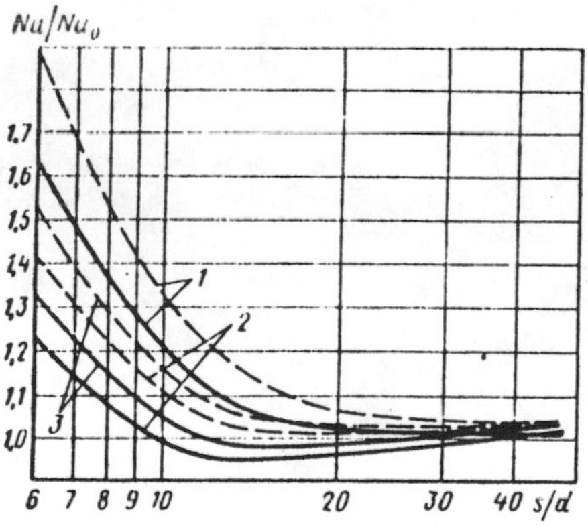

Figure 6.5 The effect of the tube twisting pitch of oval tubes on heat transfer in comparison with circular channels (—— represents T_w/T_f = 1, --- represents T_w/T_f = 3): *1)* at Re = $5\cdot 10^3$; *2)* at Re = $4\cdot 10^4$; *3)* at Re = $4\cdot 10^5$

be considerably more. As compared to a circular channel at R $= 10^5$ and $T_w/T_f = 1$, the helical tube swirling flow with $s/d = 6$ increases heat transfer by 63%, while at $T_w/T_f = 3$ it does so by 87%.

CHAPTER

SEVEN

HEAT TRANSFER AND HYDRAULIC RESISTANCE OF A CLOSE-PACKED HELICAL TUBE BUNDLE IN CROSSFLOW

Crossflow circular tube heat exchangers are widely used in different branches of industry. They are, as a rule, more effective than longitudinal-flow ones at low Reynolds numbers. Tubes with different-type fins, coiling, etc. [53] are employed to enhance heat transfer in such heat exchangers. In designing heat exchangers it is necessary to take into account not only the heat transfer efficiency but also the strength characteristics of a tube bundle, the manufacturing process for such tubes and their assembly, the possibility of choking heat exchangers in their operation, and other factors. It is important to eliminate the vibration of tube bundles in crossflow. Tube vibration may be reduced by decreasing the distance between supports or increasing the number of intermediate grids, but this results in increased metal usage and mass.

The crossflow pattern of bodies with different profiles in a cross section is rather complex. When a cylinder is in crossflow, boundary layer separation and vortex formation appear in a wake behind the boundary layer, which indicates elevated flow turbulence past the tube bundle. In this case, the flow pattern depends on the Reynolds number. The crossflow around a helical oval-shaped tube has specified features which provide a transverse velocity component along the tube axis. To estimate the flow pattern for a helical tube cross section, when the minor and major axes of an oval are parallel to the flow, it is possible, to a first approximation, to use data on the crossflow past elliptical cylinders at the same positions for the appropriate axis ratio. In [19] it was proposed that the flow past an elliptical cylinder along the minor axis has some analogy with crossflow around a plate. Crossflow past a helical tube bundle substantially differs from that around single tubes, since the flow pattern is affected by the neighboring tubes. In this case, the flow in a helical tube bundle depends on the relative tube location and the oval-shaped twisting pitches.

The application of helical oval-shaped tubes in logitudinal-flow tube bundle heat exchangers, as shown in Chapter 8, allows a 1.5-2-fold decrease in the volume of these heat exchangers as compared to circular ones. This is made possible by the improved heat transfer inside the tubes and in the intertube space of a heat exchanger. It may be expected that the circular tubes in crossflow tube bundle heat exchangers, when replaced by helical ones, also enhance heat transfer and decrease the volume of the heat exchanger. In this case, the nonuniform distribution of the heat transfer coefficient over the tube perimeter may decrease.

In a crossflow helical tube heat exchanger it is possible to implement different versions of the relative tube locations. The present chapter deals with experimental data on heat transfer and hydraulic resistance of close-packed staggered tube bundles with transverse and longitudinal pitches $a \times b$ = 12.3 x 10.6 mm. Two versions of the relative location of helical tubes in a row with twisting pitches s/d = 6.1 and 12.2 were studied. In the first version, the width of the flow channel in one row periodically varies along the tubes from zero to a maximum value equal to the difference between the maximum and minimum sizes of an oval. In this case, the neighboring tubes contact one another in each row and the tubes of adjacent rows. In each cross section, the tubes in the flow direction have the same orientation. In the second version, the tubes are mounted in each transverse row with the gaps forming, along the tube bundle, slotted channels whose width is approximately equal to half the difference between the maximum and minimum sizes of the oval. The tubes contact only those of the adjacent rows (Fig. 7.1).

7.1 EXPERIMENTAL METHODS AND MEASUREMENT DATA PROCESSING

Heat transfer and hydraulic resistance of helical oval-shaped staggered tube bundles in crossflow with air were studied in a closed aerodynamic loop [21]. The tube bundle, composed of 10

Figure 7.1 Location of helical oval-profiled tubes for a bundle with s/d = 12.2 in a row with constant width of the flow channel between the tubes

x10 rows with displacers [50], was assembled in a rectangular channel. Experiments on heat transfer were conducted under steady-state conditions at q_w = const and $T_w > T_f$. The local modelling method was used, namely, one (central) tube of row VI was heated. The tube was heated by electric d.c. current passing through its wall and was electrically insulated from the surrounding tubes at the contact points.

Heat release and tube surface temperatures were measured along the length of one twisting pitch. The inner surface temperature was measured by a travelling sensor with 4 chromel-alumel thermocouples (at the vertices of the minor and major axes of the oval) in each 1/15 of the tube twisting pitch. The arithmetic mean value of all thermocouple measurements was assumed to be a mean surface temperature. Also, the wall temperature drop was taken into account to determine the outer surface temperature.

The mean heat transfer coefficient was determined by

$$\alpha = \frac{q_w - q_{rad}}{T_w - T_f} \qquad (7.1)$$

where T_f is the flow temperature in front of a tube bundle and q_{rad} is the radiative heat flux.

The hydraulic resistance was determined through the measured pressure drop in the entire tube bundle. Experiments were made in the range of Re = 10^3–$3 \cdot 10^4$ and at a temperature factor T_w/T_f = 1.1–1.5 under steady-state conditions. When the experimental data were processed, the velocity and dimensions were taken according to the methods cited in [27] with regard to a tube bundle porosity m.

The characteristic velocity was calculated by

$$u = \frac{u_0}{m} \tag{7.2}$$

$$m = 1 - \frac{NV_{\text{tube}}}{V} \tag{7.3}$$

where u_0 is the velocity in front of a tube bundle and N is the number of tubes in a bundle.

$$d_{\text{p}} = 4\,\frac{m}{1-m}\,\frac{V_{\text{tube}}}{F_{\text{tube}}}\,\frac{\Pi_{\text{tube}}}{2S_2} \tag{7.4}$$

served as the characteristic dimension. Here, V_{tube} is the tube volume, F_{tube} the tube surface area, Π_{tube} the tube perimeter, and S_2 the longitudinal pitch of a tube grid.

Substituting into (7.4) the quantities $V_{\text{tube}} = f_{\text{area}}\, l$ and $F_{\text{tube}} = \Pi_{\text{tube}}\, l$, where f_{area} is the cross-sectional tube area determined by hydrostatic weighing and l is the tube length, we obtain:

$$d_{\text{p}} = 2\,\frac{m}{1-m}\,\frac{f_{\text{area}}}{S_2} \tag{7.5}$$

The experimental data was presented in a parametric form (Re $= \rho\, ud_{\text{p}}/\mu$, Nu $= \alpha\, d_{\text{p}}/\lambda$, Eu $= \Delta p/\rho\, u^2 z$, where z is the number of tube rows). The physical properties were evaluated at the incoming airflow and the pressure ahead of the considered row).

7.2 HYDRAULIC RESISTANCE

It might be expected that the hydraulic resistance depends on the tube length in crossflow past helical tube bundles. It is emphasized [6] that the hydraulic resistance of helical tube bundles in the crossflow is not affected by their length only if $l/0.25\ s$ is an integer. In the tube bundles investigated, this relation was equal to 10 for a tube bundle with $s/d = 6.1$, and 5 for a tube bundle with $s/d = 12.2$.

To describe the helical tube flow conditions for the first version, when the width of the flow channel along the tubes periodically changes, the experimental data on the hydraulic resistance are described well by the relations [50]:

for $s/d = 12.2$ at $\mathrm{Re} = 10^3\text{–}10^4$

$$\mathrm{Eu} = 3.86\,\mathrm{Re}^{-0.156} \tag{7.6}$$

at $\mathrm{Re} = 10^4\text{–}3\cdot10^4$

$$\mathrm{Eu} = 1.84\,\mathrm{Re}^{-0.076} \tag{7.7}$$

for $s/d = 6.1$ at $\mathrm{Re} = 10^3 - 10^4$

$$\mathrm{Eu} = 1.665\,\mathrm{Re}^{-0.060} \tag{7.8}$$

at $\mathrm{Re} = 10^4\text{–}3\cdot10^4$

$$\mathrm{Eu} = 1.18\,\mathrm{Re}^{0.021} \tag{7.9}$$

As seen in Fig. 7.2, the dependence of the hydraulic resistance on Re for the tube bundles considered is stronger at $\mathrm{Re} < 10^4$. At $\mathrm{Re} > 10^4$ this dependence noticeably weakens, and for a tube bundle with $s/d = 6.1$ it is close to the similarity relation. A comparison of the hydraulic resistances of both tube bundles investigated shows that the dependence of the hydraulic resistance on Re is weaker for the tube bundle with the smaller twisting pitch. The experimental data on the hydraulic resistance [2] for the second version of the tube location (with a constant flow channel width along the tubes) are described well by the relations:

for $s/d = 12.2$ at $\mathrm{Re} = 10^3\text{–}4.7\cdot10^3$

$$\mathrm{Eu} = 1.9 \tag{7.10}$$

at $\mathrm{Re} = 4.7\cdot10^3\text{–}3\cdot10^4$

$$\mathrm{Eu} = 8.0\,\mathrm{Re}^{-0.17} \tag{7.11}$$

for $s/d = 6.1$ at $\mathrm{Re} = 10^3\text{–}3.7\cdot10^3$

$$\mathrm{Eu} = 2.0 \tag{7.12}$$

at $\mathrm{Re} = 3.7\cdot10^3\text{–}3\cdot10^4$

$$\mathrm{Eu} = 5.78\,\mathrm{Re}^{-0.13} \tag{7.13}$$

The results indicate the weak effect of the relative twisting pitch on the Euler number.

A comparison of two versions of tube locations in a bundle shows that the hydraulic resistance of the helical tube bundles with a constant channel width between adjacent helical tubes is somewhat higher than the one with the tube bundles having a variable channel width. Figure 7.2 also illustrates the relation $\mathrm{Eu} = f(\mathrm{Re})$ obtained in [6] for a staggered helical tube bundle with $s/d = 14.2$ for 12.3 x 12.3 mm grid pitches:

$$\mathrm{Eu} = 4.65\,\mathrm{Re}^{-0.154} \tag{7.14}$$

Also shown in the figure is the relation obtained in [20] for a smooth tube bundle, for which the perimeter $\pi_{sm} = \pi_{hel}$, porosity $m_{sm} \approx m_{hel}$, and relative pitches ($a \times b = 1.23 \times 1.06$) correspond to a helical tube bundle. Unlike the data of [20], the flow velocity was determined for a smooth tube bundle not through the minimum cross section, but by Eq. (7.2). The outer tube diameter was taken as the characteristic dimension. For such a tube bundle:

at $\mathrm{Re} = 10^3\text{–}4\cdot10^3$

$$\mathrm{Eu} = 3.84\,\mathrm{Re}^{-0.15} \tag{7.15}$$

at $\mathrm{Re} = 4\cdot10^3\text{–}3\cdot10^4$

$$\mathrm{Eu} = 11.2\,\mathrm{Re}^{-0.29} \tag{7.16}$$

As seen in Fig. 7.2, only at $\mathrm{Re} > 4\cdot10^3$ is the hydraulic resistance of a smooth tube bundle smaller than that of the helical tube bundles, and at $\mathrm{Re} < 4\cdot10^3$ it practically corresponds to the tube bundle resistance when $s/d = 12.2$ with

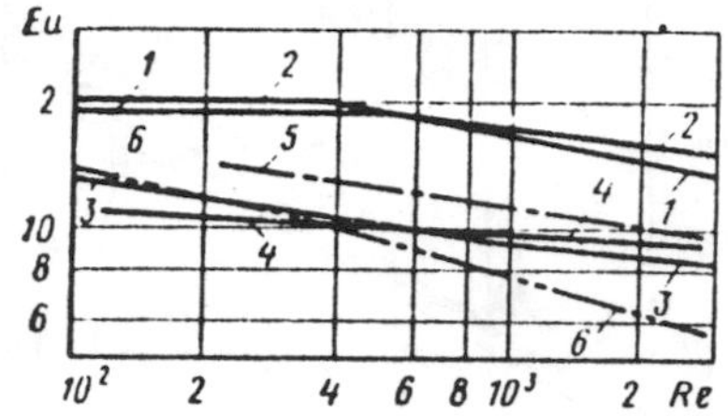

Figure 7.2 Hydraulic resistance of helical oval-profiled tube bundles: *1, 2*) for the version with a constant width of the channel between the tubes in a row with s/d = 12.2 and 6.1, respectively; *3, 4, 5*) for the version with a variable width of the channel between the tubes in a row with s/d = 12.2, 6.1, and 14.2; *6*) smooth tube bundle with $a \times b = 1.23 \times 1.06$

a variable channel width between adjacent helical tubes. The deviation observed from the data of [6] is explained by the fact that in [6] the quantity $l/0.25\ s$ is not the integer, and in this case the Euler number depends on the helical tube length.

7.3 HEAT TRANSFER

Mean heat transfer. The mean heat transfer results in crossflow past a helical tube bundle when variable width channels (Fig. 7.3) are formed between these tubes in the range of Re investigated with s/d = 6.1 are described by two relations within a limit at Re ≈ 10^4, and for a tube bundle with s/d = 12.2 by one relation [50], as in the case of smooth tube bundles:

for s/d = 6.1 at Re = 10^3–10^4

$$\mathrm{Nu} = 0.538\ \mathrm{Re}^{0.59} \tag{7.17}$$

at Re = 10^4–$3 \cdot 10^4$

$$\mathrm{Nu} = 0.232\,\mathrm{Re}^{0.88} \tag{7.18}$$

for s/d = 12.2 at Re = 10^3–$3 \cdot 10^4$

$$\mathrm{Nu} = 0.367\,\mathrm{Re}^{0.64} \tag{7.19}$$

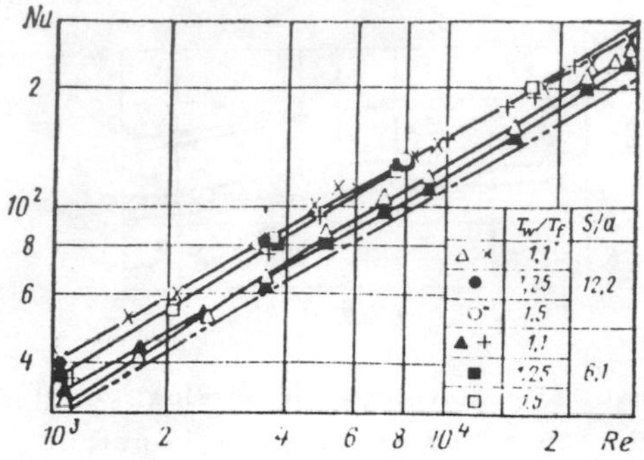

Figure 7.3 Mean heat transfer of helical tube bundles. Experimental data on heat transfer of tube bundles with a constant channel width. –··– represents a smooth tube bundle with $a \times b$ = 1.23 x 1.06

The same data presentation [27] was adopted to compare the heat transfer results in smooth circular and helical tube bundles. The pitch of a tube grid of a smooth tube bundle was equal to the helical tube pitch, and the circular tube diameter was equal to that of the helical one, i.e., the helical tub diameter was equal to the smooth tube one. The mean heat transfer from helical tube bundles with a variable flow channel between them at s/d = 6.1 and 12.2 was, on the average, 10% higher compared to circular tube bundle [20] in the Re range investigated. With increasing Re, this difference increases somewhat. In Fig. 7.3 the heat transfer coefficient for a smooth tube bundle is described by:

$$\mathrm{Nu} = 0.498\,\mathrm{Re}^{0.6} \tag{7.20}$$

The experimental data on the mean heat transfer for tube bundles, in which constant-width slotted channels exist along adjacent helical tubes [50], are described by the relations:

$$\mathrm{Nu} = 0.74\,\mathrm{Re}^{0.58} \tag{7.21}$$

$$\mathrm{Nu} = 0.51\,\mathrm{Re}^{0.62} \tag{7.22}$$

for relative twisting pitches s/d = 12.2 and 6.1 respectively. It is seen that the tube bundle heat transfer with s/d = 6.1 at Re = 10^3 is 10% less than that of a tube bundle with s/d = 12.2. At Re = 10^4 it is practically the same and at Re = $3{\cdot}10^4$ the heat transfer of a tube bundle with s/d = 6.1 is 10% higher than that of a tube bundle with s/d = 12.2. These data are, on the average, approximately 25-33% higher than those for the mean heat transfer of tube bundles with variable-width channels between adjacent tubes, and 30-40% higher than smooth tube bundles. The data point to a substantial heat transfer improvement in crossflow past a helical tube bundle for the version with slotted channels and a constant width between the tubes of a row, as compared to a smooth tube bundle and helical tube bundle with a variable channel width. The experimental data for both versions of the location of tubes in a bundle row show that the tube twisting pitch very slightly affects the mean heat transfer in crossflow past these tube bundles. No temperature factor effect on the heat transfer coefficient in the range considered was observed.

Local heat transfer. In view of the different conditions of crossflow past a helical tube bundle for the two versions investigated, differences in the distributions of the heat transfer coefficient along the tube length and perimeter should be expected. The results of the local heat transfer coefficient distributions along the perimeter and length of a tube bundle are shown in Figs. 7.4 and 7.5. Figure 7.4 illustrates distributions of the relative heat transfer coefficient over the generating lines of a helical tube passing in the middle of the lateral side of an oval and through the apex of an oval. This figure also shows positions of the tube profile and the thermocouple location relative to the incoming flow for five typical tube cross sections along the length of one pitch s. It is seen that the heat transfer both along the lateral generating line and at the apex of the oval-shaped helical tube is nonuniform. The degree of this nonuniformity does not depend on Re for the versions examined or on helical tube twisting pitches. The distribution of the local heat transfer coefficient strongly depends on the relative location of the helical tubes in a bundle (Fig. 7.4). Only one minimum of the quantity α along a length s is observed at a constant channel width between the tubes. Two minima of the quantity α along the length s are observed at a variable channel width between the tubes, which is attributed to deterioriation of the flow conditions past the tube investigated in a bundle. These occur because some part of the flow cross section is overlapped by frontal tube rows. For the tube bundle version with a variable channel width between the tubes, the nonuniformity of heat

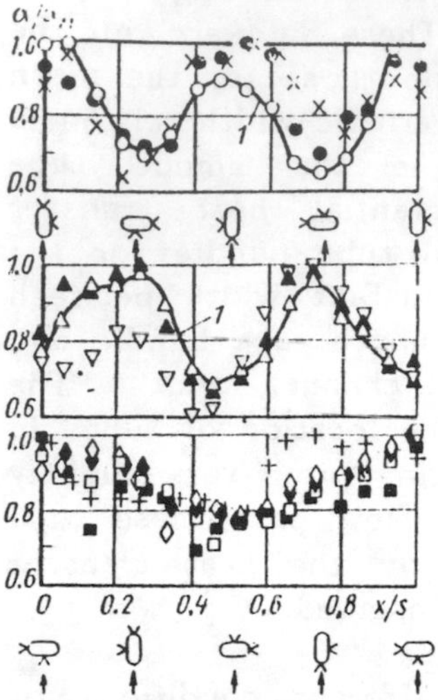

Channel width	Variable				Constant			
S/d	12,2		6,1		12,2		6,1	
	○	● ×	△	▲ ▽	□	■	◇	◆
Re	1606	21321	6770	26916	2180	28940	1109	16633
T_w/T_f	1.15	1.119	1,082	1,076	1.12	1.08	1.12	1.07
	Side	Apex	Side	Apex	Side		Side	Apex

Figure 7.4 Distribution of the relative heat transfer coefficient over the generating lines of the helical oval-shaped tube: *1*) line for varying heat transfer coefficient

transfer along the length *s* does not depend on the tube twisting pitch *s/d* and amounts to ≈ 45%, and for the tube bundle version with a constant channel width between the tubes, the nonuniformity of heat transfer is smaller and depends on *s/d*. With decreasing Re, the degree of nonuniformity of the heat transfer coefficient decreases and is 20–25% for *s/d* = 6.1 and 30–35% for *s/d* = 12.2.

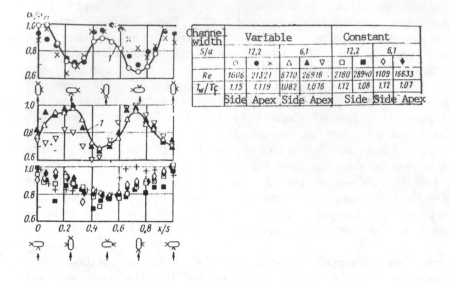

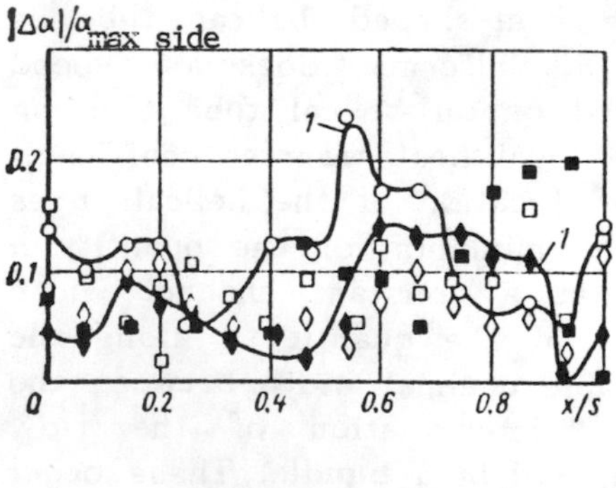

Figure 7.5 Tube length distribution of relative maximum nonuniformity of the heat transfer coefficient in the helical tube cross section. The notations are the same as in Fig. 7.4.

Experimental data on the local heat transfer coefficient for the generating line of a helical tube passing through the apex of the oval profile may also indicate that at the contact points of the tube, the heat transfer decrease is insignificant; it is higher than the one at the stern point of the lateral side of the profile. The stern part of the tube profile apex is better streamlined than the lateral side, which results in a higher heat transfer coefficient at this point. The highest heat transfer coefficient is observed at the leading critical point of the apex of the helical tube profile.

Figure 7.5 shows the perimeter change in the maximum nonuniformity of the heat transfer coefficient, i.e., the change in the absolute difference between the coefficients in the cross section of a helical tube for the apex and lateral side of the oval profile $|\Delta\alpha|$ based on the maximum heat transfer on the side surface $\alpha_{max\ side}$. It is seen from this figure that for the version of a tube bundle with a variable channel width between the tubes, the heat transfer nonuniformity over the perimeter does not depend on the twisting pitch and is about 25%, and for the tube bundle version with a constant flow channel width for a twisting pitch s/d = 12.2 it amounts to 23% and for s/d = 6.1 it decreases up to 15%. Thus, the nonuniformity of α over the helical tube perimeter is 2-3 times less than for a smooth staggered tube bundle $a \times b$ = 1.25 x 1.25 [19, 42], whereas the nonuniformity of heat transfer over the circular tube perimeter is approximately 45%.

7.4 EFFICIENCY INDICES

The energy efficiency of helical tube bundles and of a smooth tube bundle with a tube perimeter equal to that of the helical tube and with a relative pitch $a \times b$ = 1.23 x 1.06 was calculated by the formula:

$$E = \frac{\alpha}{N} \tag{7.23}$$

where N is the power spent for heat carrier pumping per unit heating surface. For such a choice of smooth tube bundle, all bundles have approximately equal heating surfaces per unit volume.

The energy efficiency of a helical tube bundle with a variable channel width between the tubes is usually less than that of a smooth tube bundle [50]. For a tube bundle with a twisting pitch s/d = 6.1, only at N < 1 W/m^2 is the energy efficiency slightly higher than that of a smooth tube bundle, and for a tube bundle with s/d = 12.2 in the Re range investigated the energy efficiency, as compared to that of a smooth tube bundle, is, on the average, less than ≈ 11%. The difference increases with increasing Re. At Re < 10^4 the bundles of strongly twisted tubes are more effective, and at Re > 10^4 the tube bundles with a constant channel width between the tubes at the beginning of the Re range considered, the energy efficiency of a tube bundle with s/d = 6.1 (at N = 0.5 W/m^2) exceeds that of a smooth tube by 4%, and that of a tube bundle with s/d = 12.2 by 11%. The energy efficiency of a tube bundle with s/d = 6.1 is the same as in the case of a smooth tube bundle at $N \approx 32$ W/m^2 (Re ≈ $3.7 \cdot 10^3$), as is that of a tube bundle with s/d = 12.2 at $N \approx 80$ W/m^2 (Re ≈ $5.2 \cdot 10^3$). At large N the efficiency of a smooth tube bundle is higher. Thus, at $N = 10^4$ W/m^2 the efficiency of a smooth tube bundle is 4% higher than a tube bundle with s/d = 6.1, and 9% with s/d = 12.2. At N < 80 W/m^2 (≈ Re $5.2 \cdot 10^3$) the tube bundles with a constant channel width between the tubes and with s/d = 12.2 are most effective.

CHAPTER
EIGHT

THE ENGINEERING METHODS OF THERMAL AND HYDRAULIC CALCULATION OF CLOSE-PACKED HELICAL TUBE HEAT EXCHANGERS

The design relations presented in Chapters 5–7 for heat transfer and hydraulic resistance of close-packed helical tube bundles in longitudinal flow, of helical tube bundles in crossflow, and with heat carriers flowing inside helical tubes make it possible, using ordinary methods, to perform thermal and hydraulic calculations of heat exchangers composed of helical tubes. However, choosing an optimum helical tube twisting pitch should precede such calculations. Therefore, the present chapter proposes methods of estimating the efficiency of heat transfer enhancement in helical tube heat exchangers.

Usually, in engineering practice, when a heat exchanger is being designed a thermal calculation is first made. The results of this thermal calculation allow determining the heat exchanger size. Then, a hydraulic calculation is made. If the hydraulic losses of a heat exchanger do not agree with the initial specifications, then subsequent versions are calculated by assigning different flow velocities. In spite of the fact that calculations are performed by computer, these methods are tedious and hamper the optimization of heat exchanger parameters. Therefore, the present chapter deals with methods of heat exchanger calculation that allow a direct determination of the sizes of the heat exchanger at assigned values of its heat power and hydraulic resistances on the hot and cold sides.

8.1 ESTIMATING THE EFFICIENCY OF HEAT TRANSFER ENHANCEMENT IN HEAT EXCHANGERS WITH TUBE BUNDLES IN LONGITUDINAL FLOW

Different methods of heat transfer improvement are widely used in designing high-efficiency heat exchangers. This enables decreasing heat exchanger sizes and mass, either to increase the heat power of the exchanger or to solve other specific problems.

Rather simple procedures are available to compare the efficiency of heat exchangers and to choose both a proper method of heat transfer enhancement and optimum parameters of the exchangers. Such a method has been proposed in [26] for tube heat exchangers operating under both fully developed turbulent flow and arbitrary flow conditions. However, in both cases the assumption was made that the compared channels in which the heat transfer occurs had the same equivalent diameters.

Usually, the efficiency of heat transfer improvement may be estimated by several criteria.

1. A comparison of the heating surfaces or volumes of two heat exchangers: one with smooth surfaces and another with devices to enhance the heat transfer. In this case, both heat exchangers have the same heat power, heat carrier flowrates, and pressure losses for their pumping, i.e., these are characterized by the same pumping powers.

2. A comparison of heat powers of heat exchangers having the same volumes, heat carrier flowrates, and pressure losses for heat carrier pumping (heat power pumping losses) at the same volumes, heat power, and flowrates of the heat carriers. In this comparison it is assumed that the compared channels have the same perimeters, i.e.,

$$\Pi = \Pi_{sm} \tag{8.1}$$

The subscript "sm" refers to a smooth channel.

First, let us obtain a comparison criterion for heating surfaces or heat exchanger volumes under the same heat power, heat carrier flowrates in tubes, and hydraulic resistance, i.e.,

$$Q = Q_{sm} \tag{8.2}$$

$$G = G_{sm} \tag{8.3}$$

$$\Delta p = \Delta p_{sm} \tag{8.4}$$

We consider that the thermal resistance on one side is small. As $Q = \alpha \Delta T \Pi l N$ and $Q_{sm} = \alpha T_{sm} l_{sm} \Pi_{sm} N_{sm}$, where α is the heat transfer coefficient, ΔT the appropriate temperature difference, l the tube length, N the number of tubes in a heat exchanger, and the temperature heads of both heat exchangers are equal ($\Delta T = \Delta T_{sm}$), then according to (8.2) we have:

$$\frac{\alpha l N}{\alpha_{sm} l_{sm} N_{sm}} = l \tag{8.5}$$

Since the friction pressure losses are

$$\Delta p = \xi \frac{l}{d_{eq}} \frac{\rho u^2}{2} \quad \text{and} \quad \Delta p_{sm} = \frac{l_{sm}}{l_{eq.sm}} \frac{\rho_{sm} u_{sm}^2}{2}$$

where u is the mean velocity in the tubes, ρ the density, ξ the hydraulic resistance coefficient, and $\rho = \rho_{sm}$, according to (8.4)

$$\frac{\xi}{\xi_{sm}} \frac{l}{l_{sm}} \frac{d_{eq.sm}}{d_{eq}} \frac{u^2}{u_{sm}^2} = 1 \tag{8.6}$$

From, Eq. (8.3) it follows that the Reynolds numbers for the channels are related by:

$$\frac{Re}{Re_{sm}} = \frac{u d_{eq}}{u_{sm} d_{eq.sm}} = \frac{f_{sm} N_{sm} d_{eq}}{f N d_{eq.sm}} = \frac{N_{sm}}{N} \tag{8.7}$$

Here, f is the area of the flow cross section of the channel.

For smooth channels, $Nu_{sm} = C_1 Re^n$ and $\xi_{sm} = C_2 Re^m$. For channels with heat transfer enhancement, the heat transfer and pressure loss increases are accounted for by the ratios Nu/Nu_{sm} and ξ/ξ_{sm}, which for the prescribed channel geometry depend on Re. Hence,

$$Nu = \left(\frac{Nu}{Nu_{sm}}\right)_{Re} C_1 Re^n; \quad \xi = \left(\frac{\xi}{\xi_{sm}}\right)_{Re} C_2 Re^m$$

$$\frac{\alpha}{\alpha_{sm}} = \left(\frac{Nu}{Nu_{sm}}\right)_{Re} \left(\frac{Re}{Re_{sm}}\right)^n \frac{d_{eq.sm}}{d_{eq}} \tag{8.8}$$

$$\frac{\xi}{\xi_{sm}} = \left(\frac{\xi}{\xi_{sm}}\right)_{Re} \left(\frac{Re}{Re_{sm}}\right)^m \tag{8.9}$$

where the subscript "Re" means that the ratios Nu/Nu_{sm} and ξ/ξ_{sm} are taken at the same Reynolds numbers for the smooth channel and the channel with heat transfer enhancement, which are equal to the Reynolds number for the channel with heat transfer enhancement.

From Eqs. (8.5) and (8.9) it is possible to obtain the tube number ratio as well as the tube length ratio for the compared heat exchangers

$$\frac{N}{N_{sm}} = \frac{\left(\frac{\xi}{\xi_{sm}}\right)_{Re}^{\frac{1}{3-n+m}}}{\left(\frac{Nu}{Nu_{sm}}\right)_{Re}^{\frac{1}{3-n+m}} \left(\frac{d_{eq}}{d_{eq.sm}}\right)^{\frac{2}{3-n+m}}} \tag{8.10}$$

$$\frac{l}{l_{sm}} = \frac{\left(\frac{d_{eq}}{d_{eq.sm}}\right)^{\frac{5-3n+m}{3-n+m}}}{\left(\frac{Nu}{Nu_{sm}}\right)_{Re}^{\frac{2+m}{3-n+m}} \left(\frac{\xi}{\xi_{sm}}\right)_{Re}^{\frac{1}{3-n+m}}} \tag{8.11}$$

If the compared tubes are located in a bundle with the same pitch, then the heat exchanger cross-sectional area ratio is

$$\frac{F}{F_{sm}} = \frac{N}{N_{sm}} = \frac{\left(\frac{\xi}{\xi_{sm}}\right)_{Re}^{\frac{1}{3-n+m}}}{\left(\frac{Nu}{Nu_{sm}}\right)_{Re}^{\frac{1}{3-n+m}} \left(\frac{deq}{d_{eq.sm}}\right)^{\frac{2}{3-n+m}}} \tag{8.12}$$

The heat exchanger volume ratio is

$$K_v = \frac{V}{V_{sm}} = \frac{Fl}{F_{sm} l_{sm}} = \frac{\left(\frac{\xi}{\xi_{sm}}\right)_{Re}^{\frac{n}{3-n+m}} \left(\frac{d_{eq}}{d_{eq.sm}}\right)^{\frac{3-3n+m}{3-n+m}}}{\left(\frac{Nu}{Nu_{sm}}\right)_{Re}^{\frac{3+m}{3-n+m}}} \tag{8.13}$$

Under the turbulent flow conditions $n = 0.8$ and $m = -0.2$, and thus we have:

$$K_v = \frac{\left(\frac{\xi}{\xi_{sm}}\right)_{Re}^{0.4} \left(\frac{deq}{deq.sm}\right)^{0.2}}{(Nu/Nu_{sm})_{Re}^{1.4}} \tag{8.14}$$

The heat powers of the heat exchangers are to be compared at the same heat carrier flowrates, pressure losses, and heat exchanger volumes. Since the compared heat exchangers are composed of tubes having the same perimeter and are located in a bundle with the same pitch, they have the same heating surfaces, i.e.,

$$\Pi l N = \Pi_{sm}\, l_{sm}\, N_{sm}$$

hence, it follows that

$$\frac{l}{l_{sm}} = \frac{N_{sm}}{N} \tag{8.15}$$

and, also,

$$Q/Q_{sm} = \alpha/\alpha_{sm} \tag{8.16}$$

if, as in the first case, $\Delta T = \Delta T_{sm}$. Considering Eqs. (8.6)-(8.9) and (8.16), we obtain

$$K_Q = \frac{Q}{Q_{sm}} = \frac{\left(\frac{\mathrm{Nu}}{\mathrm{Nu}_{sm}}\right)_{\mathrm{Re}} \left(\frac{d_{eq}}{d_{eq.sm}}\right)^{\frac{3n-m-1}{m+3}}}{(\xi/\xi_{sm})^{\frac{n}{3+m}}} \tag{8.17}$$

Since under turbulent flow conditions (n = 0.8 and m = -0.2) we have:

$$K_Q = \frac{\left(\frac{\mathrm{Nu}}{\mathrm{Nu}_{sm}}\right)_{\mathrm{Re}} \left(\frac{d_{eq}}{d_{eq.sm}}\right)^{-0.113}}{(\xi/\xi_{sm})^{0.286}_{\mathrm{Re}}} \tag{8.18}$$

Finally, let us compare heat exchangers with respect to pressure losses spent for heat carrier pumping. In this case, the heat carrier flowrates, volumes, and heat powers are assumed to be the same. Hence, it follows that at the same temperature differences we arrive at:

$$\alpha = \alpha_{sm} \tag{8.19}$$

From Eqs. (8.7)-(8.9) and (8.19) we obtain

$$K_{\Delta p} = \frac{\Delta p}{\Delta p_{sm}} = \frac{\left(\frac{\xi}{\xi_{sm}}\right)_{\mathrm{Re}} \left(\frac{deq}{deq.sm}\right)^{\frac{3+m-3n}{n}}}{(\mathrm{Nu}/\mathrm{Nu}_{sm})_{\mathrm{Re}}^{\frac{3+m}{n}}} \tag{8.20}$$

Under turbulent flow conditions (n = 0.8 and m =-0.2) we have:

$$K_{\Delta p} = \frac{\left(\frac{\xi}{\xi_{sm}}\right)_{\mathrm{Re}} \left(\frac{d_{eq}}{deq.sm}\right)^{0.5}}{(\mathrm{Nu}/\mathrm{Nu}_{sm})^{3.5}_{\mathrm{Re}}} \tag{8.21}$$

In these calculations the heat transfer of a heating surface was considered but the thermal resistance was not taken into account. If the relations K_{V1} and K_{V2} are known for both sides of a heat exchanger, then, ignoring the thermal resistance of the tube walls, the expression for the volume ratio of the compared heat exchangers may be obtained:

$$K_V = \frac{V}{V_{sm}} = K_{V1} \left[\frac{1 + \alpha_1/\alpha_2}{1 + \dfrac{K_{V1}}{K_{V2}} \dfrac{\alpha_1}{\alpha_2}} \right] \tag{8.22}$$

where α_1 and α_2 are the heat transfer coefficients for the inner and outer tube surfaces with heat transfer enhancement,

$$\frac{\alpha_1}{\alpha_2} = \frac{\alpha_1}{\alpha_{1sm}} \frac{\alpha_{2sm}}{\alpha_2} \frac{\alpha_{1sm}}{\alpha_{2sm}} = \left(\frac{\text{Nu}}{\text{Nu}_{sm}}\right)_1 \left(\frac{\text{Nu}_{sm}}{\text{Nu}}\right)_2 \frac{\alpha_{1sm}}{\alpha_{2sm}} \tag{8.23}$$

Since for the compared surfaces, the ratios $(\text{Nu}/\text{Nu}_{sm})_1$ and $(\text{Nu}/\text{Nu}_{sm})_2$ are known, the value of K_V may be found by assigning the heat transfer coefficient ratio $\alpha_{1sm}/\alpha_{2sm}$ for a smooth tube heat exchanger.

The above relations allow a simple comparison of the efficiency of heat transfer augmentation if the compared channels have different equivalent diameters.

8.2 ESTIMATING THE EFFICIENCY OF HEAT TRANSFER IMPROVEMENT IN CROSSFLOW TUBE BUNDLE HEAT EXCHANGERS

Crossflow tube bundle heat exchangers are widely used in different branches of technology.

Let us consider a heat exchanger whose heat transfer coefficient inside the tubes is substantially higher than that on the outside. The tube bundles to be compared have the same longitudinal and transverse tube pitches (S_1 and S_2, Fig. 8.1). It is also assumed that the bundles consist of tubes with the same perimeter, i.e.,

$$\Pi = \Pi_{sm} \tag{8.24}$$

The subscript "sm" refers to a smooth tube bundle.

As a result, we obtain a criterion for comparing heating surfaces or volumes of heat exchangers with the same heat power, heat carrier flowrate, and hydraulic resistance, i.e.,

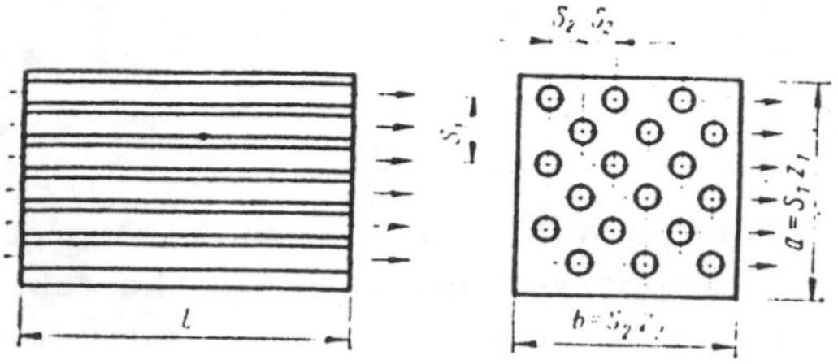

Figure 8.1 Schematic of a heat exchanger

$$Q = Q_{sm} \tag{8.25}$$

$$G + G_{sm} \tag{8.26}$$

$$\Delta p = \Delta p_{sm} \tag{8.27}$$

with $Q = \alpha \Delta T l\, z z_1$ and $Q_{sm} = \alpha_{sm}\, \Delta T_{sm}\, \Pi_{sm}\, l_{sm}\, z_{sm}\, z_{1sm}$, where α is the heat transfer coefficient in the intertube space, ΔT the temperature difference, l the tube length, z the number of tube rows in the longitudinal direction, z_1 the number of tube rows in the transverse direction. Since the temperature differences for both heat exchangers are identical ($\Delta T = \Delta T_{sm}$), according to (8.25) we have:

$$\frac{\alpha l z_1 z}{\alpha_{sm} l_{sm} z_{1sm} z_{sm}} = 1 \tag{8.28}$$

Since the pressure losses in crossflow past a tube bundle are

$$\Delta p = \xi z \frac{\rho u^2}{2} \text{ and } \Delta p_{sm} = \xi_{sm} z_{sm} \frac{\rho_{sm} u_{sm}^2}{2} \tag{8.29}$$

where u is the characteristic velocity, ρ is the density, ξ is the resistance coefficient, and $\rho = \rho_{sm}$, according to (8.27) we arrive at:

$$\frac{\xi}{\xi_{sm}} \frac{z}{z_{sm}} \frac{u^2}{u_{sm}^2} = 1 \tag{8.30}$$

For tube bundles of a noncircular cross section in crossflow, use is made of the characteristic dimension proposed in [27]

$$d_p = 4\frac{\psi}{1-\psi}\;\frac{V}{F}\;\frac{L'}{s_2} \tag{8.31}$$

where ψ is the tube bundle porosity with respect to a heat carrier, V_{tube} is the tube volume, F_{tube} is the tube surface, L' is the half tube perimeter, and s_2 is the longitudinal pitch.

For the smooth tube bundle with $s_1/d = 1.23$ and $s_2/d_2 = 1.065$, the characteristic dimension is equal to the outer tube diameter, i.e.,

$$d_{p.sm} = d_{sm} \tag{8.32}$$

Since $V_{tube} = f_{area}$, $F_{tube} = \Pi l$ and $L' = \Pi/2$, where f_{area} is the cross-sectional tube area, we have:

$$d_p = \frac{2\psi}{1-\psi}\;\frac{f_{area}}{s_2} \tag{8.33}$$

The characteristic velocity is

$$u = u_0/\psi \tag{8.34}$$

Therefore, $G = \rho u \psi l \alpha z_1$ and $Q_{sm} = \rho_{sm} u_{sm} \psi_{sm} l_{sm} \alpha_{sm} z_{1sm}$, and from (8.26) it follows that

$$\frac{u \varphi l z_1}{u_{sm} \psi_{sm} l_{sm} z_{1sm}} = 1 \tag{8.35}$$

The Reynolds number ratio for the heat exchangers is

$$\frac{Re}{Re_{sm}} = \frac{u d_p}{u_{sm} d_{p.sm}} = \frac{l_{sm} z_{1sm} (1-\psi_{sm}) f_{area}}{l z_1 (1-\psi)\; f_{area.sm}} \tag{8.36}$$

For smooth tube bundles

$$Nu_{sm} = C_1\, Re_{sm}^{n} \quad \text{and} \quad \xi_{sm} = C_2\, Re_{sm}^{m}$$

For tube bundles with heat transfer enhancement, an increase in heat transfer and hydraulic resistance is taken into consideration by the ratios Nu/Nu_{sm} and ξ/ξ_{sm} that depend on Re when the tube bundle geometry is predetermined. Therefore,

$$Nu=\left(\frac{Nu}{Nu_{sm}}\right)_{Re} C_1 Re^n;\ \xi=\left(\frac{\xi}{\xi_{sm}}\right)_{Re} C_2 Re^m$$

$$\frac{\alpha}{\alpha_{sm}}=\left(\frac{Nu}{Nu_{sm}}\right)_{Re}\left(\frac{Re}{Re_{sm}}\right)^n \frac{d_{\mu sm}}{d_\mu} \tag{8.37}$$

$$\frac{\xi}{\xi_{sm}}=\left(\frac{\xi}{\xi_{sm}}\right)_{Re}\left(\frac{Re}{Re_{sm}}\right)^m \tag{8.38}$$

where the subscript "Re" means that the ratios Nu/Nu_{sm} and ξ/ξ_{sm} are taken at the same Reynolds numbers for bundles composed of smooth tubes and tubes with heat transfer enhancement. In this case, the Reynolds number for a helical tube is considered.

For Eqs. (8.28)-(8.38) we may obtain the ratios of the cross-sectional areas and of the number of longitudinal tube rows for the compared heat exchangers:

$$\frac{lz_1}{l_{sm}z_{1sm}}=\left[\frac{\left(\frac{\xi}{\xi_{sm}}\right)_{Re}\left(\frac{1-\psi}{1-\psi_{sm}}\right)^{n-m-1}\left(\frac{f_{area}}{f_{area.sm}}\right)^{m-n+1}}{(Nu/Nu_{sm})_{Re}\,(\psi/\psi_{sm})}\right]^{\frac{1}{3+m-n}} \tag{8.39}$$

$$\frac{z}{z_{sm}}=\left[\frac{\left(\frac{1-\psi}{1-\psi_{sm}}\right)^{2n-2}\left(\frac{\psi}{\psi_{sm}}\right)^{1+m-2n}\left(\frac{f_{area}}{f_{area.sm}}\right)^{2-2n}}{(Nu/Nu_{sm})_{Re}^{m+2}\,(\xi/\xi_{sm})_{Re}^{1-n}}\right]^{\frac{1}{3+m-n}} \tag{8.40}$$

The heat exchanger volume ratio is

$$K_{V1}=V/V_{sm}=lz_1z/l_{sm}z_{1sm}z_{sm}=$$

$$\left[\frac{\left(\frac{\xi}{\xi_{sm}}\right)_{Re}^{n}\left(\frac{\psi}{\psi_{sm}}\right)^{3+m-2n}\left(\frac{1-\psi}{1-\psi_{sm}}\right)^{3n-3-m}\left(\frac{f_{area}}{f_{area.sm}}\right)^{3+m-3n}}{(Nu/Nu_{sm})_{Re}^{m+3}}\right]^{\frac{1}{3+m-n}} \tag{8.41}$$

For tubes with a constant cross-sectional area we have

$$\frac{1-\psi}{1-\psi_{sm}}=\frac{f_{area}}{f_{area.sm}} \tag{8.42}$$

because the compared tube bundles have the same pitch. Therefore,

$$\frac{l z_1}{l_{sm} z_{1sm}} = \left[\frac{\left(\dfrac{\xi}{\xi_{sm}}\right)_{Re}}{\left(\dfrac{Nu}{Nu_{sm}}\right)_{Re} \left(\dfrac{\psi}{\psi_{sm}}\right)} \right]^{\frac{1}{3+m-n}} \tag{8.43}$$

$$\frac{z}{z_{sm}} = \left[\frac{\left(\dfrac{\psi}{\psi_{sm}}\right)^{4+m-2n}}{\left(\dfrac{Nu}{Nu_{sm}}\right)_{Re}^{m+2} \left(\dfrac{\xi}{\xi_{sm}}\right)_{Re}^{1-n}} \right]^{\frac{1}{3+m-n}} \tag{8.44}$$

$$K_{V1} = \frac{V}{V_{sm}} = \left[\frac{\left(\dfrac{\xi}{\xi_{sm}}\right)_{Re}^{n} \left(\dfrac{\psi}{\psi_{sm}}\right)^{3+m-2n}}{(Nu/Nu_{sm})_{Re}^{m+3}} \right]^{\frac{1}{3+m-n}} \tag{8.45}$$

The above relations allow a determination of the volume ratio at small thermal resistance inside the tubes. If, vice versa, the thermal resistance on the outside of the tubes may be neglected, then according to section 8.1 we have:

$$K_{V2} = \frac{V}{V_{sm}} = \left[\frac{\left(\dfrac{\xi}{\xi_{sm}}\right)_{Re}^{n} \left(\dfrac{d_{eq}}{d_{eq.sm}}\right)^{3-3n+m}}{(Nu/Nu_{sm})_{Re}^{3+m}} \right]^{\frac{1}{3+m-n}} \tag{8.46}$$

where d_{eq} is the equivalent diameter of the inner channel. If the relations K_{V_1} and K_{V_2} are known for both sides of a heat exchanger, then, when the thermal resistance of the tube walls is neglected, the volume ratio of the heat exchangers is found by Eqs. (8.22) and (8.23).

For the compared surfaces, $(Nu/Nu_{sm})_1$ and $(Nu/Nu_{sm})_2$ are known. K_V may be determined by assigning heat transfer coefficient ratios, $\alpha_{1sm}/\alpha_{2sm}$, for a smooth tube heat exchanger.

8.3 CALCULATION METHODS FOR HEAT EXCHANGERS AT A GIVEN HYDRAULIC RESISTANCE

As mentioned, the relationships obtained for heat transfer and hydraulic resistance in the heat carrier flow on the outside and inside of helical tubes may be used to calculate the characteristics of specific heat exchangers with such tubes. The calculation methods usually adopted [19, 38, 39] assume that the flow velocity is prescribed and the heating surface may be determined at the specific heat exchanger heat power and heat carrier flowrates. The hydraulic calculation is aimed at determining whether the hydraulic resistance of a heat exchanger corresponds to the specified value. In the case of noncoincidence, the calculation is repeated at other flow velocities. It is obvious that the relationship between flow velocity and hydraulic losses is strong and may be employed to determine a heating surface and heat exchanger size at once, without several approximations. The appropriate methods are proposed below.

The heat exchanger with a tube bundle in longitudinal flow. The initial data are an inlet hot heat carrier temperature T'_h, an outlet hot heat carrier temperature T''_h, its flowrate G_h, an inlet pressure p'_h, hydraulic resistance Δp_h, an inlet cold heat carrier temperature T'_c, an outlet cold heat carrier temperature T''_c, its flowrate G_c, an inlet pressure p'_c, and hydraulic resistance Δp_c. The geometrical parameters of the helical tubes are also prescribed. These tubes form a bundle. The characteristic dimensions $d_{eq.in}$ and $d_{eq.out}$ are the equivalent inner and outer tube diameters; Π_{in} and Π_{out} are the wetted inner and outer perimeters; S_{in} and S_{out} are the inner and outer areas of the flow cross sections (part of the intertube space per one tube; s is the tube twisting pitch; and d_{in} and d_{out} are the maximum inner and outer sizes of an oval.

First, the exchanger heat power Q, mean logarithmic temperature difference ΔT_{log}, heat carrier mean temperatures $\bar{T}_h$ and $\bar{T}_c$, and the mean heat carrier densities $\bar{\rho}_h$ and $\bar{\rho}_c$ are determined from the specified data.

Let us consider the hot heat carrier as flowing inside the tubes while the cold one flows in the intertube space, and both heat carriers are gaseous.

Let us examine the tube flow. The pressure losses in a gas flow are those spent for friction, local resistances, and flow acceleration. When the variation in the gas velocity is neglected in calculations of local resistances, the pressure losses are expressed by the relation:

$$\Delta p_h = \left[\xi_h \frac{l}{d_{eq.in}} m_h + \Sigma\xi_i + 2\frac{T''_h - T'_h}{\bar{T}_h} + 4\frac{1-\sigma_h}{1+\sigma_h}\right]\frac{\bar{\rho}_h \bar{u}_h^2}{2} \tag{8.47}$$

where ξ_h is the hydraulic resistance coefficient, l the tube length, m_h the number of passes by the hot gas, ξ_i the local resistance coefficient, $\bar{u}_h$ the mean gas velocity in the tube, and $\sigma_h = 1 - \Delta p_h/p_h/p'_h$.

In [36], Eq. (8.47) was reduced to the form:

$$\sigma_h = \sqrt{1 - \frac{\mathrm{Re}_h^2(1+\sigma_h)\mu_h^2}{4\bar{\rho}_h^2 d_{\mathrm{eq.in}}^2 R_h \bar{T}_h}\left[\xi_h \frac{l}{d_{\mathrm{eq.in}}} m_h + \Sigma\xi_i + 2\,\frac{T_h'' - T_h'}{\bar{T}_h} + 4\,\frac{1-\sigma_h}{1+\sigma_h}\right]} \tag{8.48}$$

where μ_h is the dynamic viscosity, Re_h is the Reynolds number on the hot side, and R_h is the gas constant. From Eq. (8.48) we may obtain:

$$\mathrm{Re}_h^2 = \frac{4(1-\sigma_h^2)\,\rho_h^2 d_{\mathrm{eq.in}}^2 R\,\bar{T}_h}{(1+\sigma_h)^2\mu_h^2\left[\xi_h \frac{l}{d_{\mathrm{eq.in}}} m_h + \Sigma\xi_i + \frac{2\left(T_h'' - T_h'\right)}{\bar{T}_h} + 4\,\frac{1-\sigma_h}{1+\sigma_h}\right]} \tag{8.49}$$

In Eq. (8.49), the product

$$\xi_h \frac{l}{d_{\mathrm{eq.in}}} m_h$$

is the only unknown quantity and weakly depends on the Reynolds number. For helical tubes, with heat transfer enhancement, we have:

$$\xi_h = 0.3164/\mathrm{Re}_h^{0.25}\left(\frac{\xi_h}{\xi_{h.sm}}\right)\mathrm{Re}_h \tag{8.50}$$

Equation (8.50) is valid for turbulent flow under isothermal flow conditions as well as for gas cooling.

Let us designate the area of the flow cross section of one pass through S_h; the heating surface, determined by the inner tube perimeter, F_h; and the total number of tubes, N. Since

$$l = F_h/\Pi_{\mathrm{in}} N \tag{8.51}$$

$$N = \frac{4 S_h m_h}{\Pi_{\mathrm{in}} d_{\mathrm{eq.in}}} \tag{8.52}$$

it may be stated that

$$\frac{l}{d_{eq.in}} = \frac{F_h}{4S_h\, m_h} \tag{8.53}$$

The heat flux transferred through a heating surface is

$$Q = \alpha_h\, F_h\, \Delta T_h \tag{8.54}$$

where ΔT_h is the mean temperature difference on the hot side. On the other hand,

$$Q = G_h c_{ph}(T'_h - T''_h) = \bar{u}_h \bar{\rho}_h S_h c_{ph}(T'_h - T''_h) \tag{8.55}$$

where c_{ph} is the mean heat capacity. Using Eqs. (8.53)-(8.55), we arrive at:

$$\frac{l}{d_{eq.in}}\, m_h = \frac{\bar{u}_h \bar{\rho}_h c_{ph}(T'_h - T''_h)}{4\alpha_h \Delta T_h} = \frac{Re\, \mu_h c_{ph}(T'_h - T''_h)}{4 d_{eq.in}\, \alpha_h \Delta T_h} \tag{8.56}$$

For turbulent gas flow in a helical tube with gas cooling, the heat transfer coefficient may be expressed as:

$$\alpha_h = \left(\frac{\mathrm{Nu}_h}{\mathrm{Nu}_{h.sm}}\right)_{\mathrm{Re}_h} \frac{0.023\, \mathrm{Re}_h^{0.8}\, \mathrm{Pr}_h^{0.4}\, \lambda_h}{d_{eq.in}} = K_h\, \mathrm{Re}_h^{0.8} \tag{8.57}$$

where λ_h is the thermal conductivity.

Substituting Eq. (8.57) into (8.56) and using the definition of the Prandtl number $\mathrm{Pr}_h = \mu_h\, c_{ph}/\lambda_h$ gives:

$$\frac{l}{d_{eq.in}}\, m_h = \frac{\mathrm{Re}_h^{0.2}\, \mathrm{Pr}_h^{0.6}\, (T'_h - T''_h)}{0.092 \Delta T_h (\mathrm{Nu}_h \mathrm{Nu}_{h.sm})_{\mathrm{Re}_h}} \tag{8.58}$$

Using (8.50), we have:

$$\xi_h \frac{l}{d_{eq.in}}\, m_h = \frac{3.44 \mathrm{Pr}_h^{0.6}\, (T'_h - T''_h)}{\mathrm{Re}_h^{0.05}\, \Delta T_h} \, \frac{(\xi_h/\xi_{h.sm})_{\mathrm{Re}_h}}{(\mathrm{Nu}_h/\mathrm{Nu}_{h.sm})_{\mathrm{Re}_h}}$$

$$= \frac{1.75(T'_h - T''_h)}{\Delta T_h} \frac{(\xi_h/\xi_{h.sm})_{Re_h}}{(Nu_h/Nu_{h.sm})_{Re_h}} = \frac{K_1}{\Delta T_h} \tag{8.59}$$

(for gases $Pr_h = 0.71$, and for $Re_h = 10^4$–10^5 it is assumed that $Re_{hot}^{0.05} = 1.6$). If the temperature difference in the tube wall is neglected, then the following expression is obtained:

$$\Delta T_h = \frac{d_{out}/d_{in}}{N_\alpha + d_{out}/d_{in}} \Delta T_{log} \tag{8.60}$$

where $N_\alpha = \alpha_h/\alpha_c$ is the relationship between the heat transfer coefficients on the hot and cold sides.

Now let us consider flow in the intertube space. As shown in Chapter 5, the equations for calculating the coefficients of hydraulic resistance and heat transfer in turbulent flow may be reduced to the form:

$$\xi_c = B\xi_{tube} = (1 + 3.6\ Fr_m^{-0.357}) \frac{0.3164}{Re_c^{0.25}} = \Phi_1(s/d) \frac{0.3164}{Re_c^{0.25}}$$

$$= \left(\frac{\xi_c}{\xi_{c.sm}}\right)_{Re_c} \frac{0.3164}{Re_c^{0.25}} \tag{8.61}$$

$$\alpha_c = 0.023\ Re_c^{0.8}\ Pr_c^{0.4} \left[1 + \frac{3.6}{Fr_m^{0.357}}\right](T_w/T_c)^{-0.55} \frac{\lambda_c}{d_{eq.out}}$$

$$= \Phi_2(s/d)\ K_c\ Re_c^{0.8} = (Nu_c/Nu_{c.sm})_{Re_c}\ K_c\ Re_c^{0.8} \tag{8.62}$$

Repeating the above calculations for the case of tube bundle flow and using Eqs. (8.61) and (8.62), we arrive at:

$$Re_c^2 = \frac{4(1-\sigma_c^2)\bar{\rho}_c^2 d_{eq.out}^2 R_c \bar{T}_c}{(1+\sigma_c)^2\mu_c^2\left[\xi_c \frac{l}{d_{eq.out}} m_c + \Sigma\xi_i + 2\frac{T''_c - T'_c}{\bar{T}_c} + 4\frac{1-\sigma_c}{1+\sigma_c}\right]} \tag{8.63}$$

$$\xi_c \frac{l}{d_{eq.out}} m_c = \frac{3.44\ Pr_c^{0.8}(T''_c - T'_c)}{Re^{0.05}\Delta T_c} - \left(\frac{T_w}{T_c}\right)^{0.55} \frac{(\xi_c/\xi_{c.sm})_{Re_c}}{(Nu_c/Nu_{c.sm})_{Re_c}}$$

$$= \frac{1.75(T''_c - T'_c)}{\Delta T_c} - \left(\frac{T_w}{T_c}\right)^{0.55} \frac{(\xi_c/\xi_{c.sm})_{Re_c}}{(Nu_c/Nu_{c.sm})_{Re_c}} \tag{8.64}$$

Here, $\sigma_c = 1\ (p_c/\rho'_c)$, R_c is the gas constant for the cold fluid, μ_c is the dynamic viscosity for the cold fluid, m_c is the number of passes of the cold fluid, and ΔT_c is the mean temperature difference on the cold side. When $\xi_c(l/d_{eq.out})m_c$ is determined by Eq. (8.64), the wall temperature may be estimated or it may be considered approximately that $(T_w/T_c)^{0.55} \approx 1$. If $\Phi_1(s/d) = \Phi_2(s/d)$, which was observed at large values of s/d, then Eq. (8.64) can be simplified.

The quantity ΔT_c is determined by the expression:

$$\Delta T_c = \frac{N_\alpha}{N_\alpha + d_{in}/d_{out}} \Delta \bar{T}_{log} \tag{8.65}$$

The heat transfer coefficient ratio N_α depends on the relationship between the flowrates of the hot and cold fluids. If $m_c = m_h$, then using Eq. (8.57) and (8.62) as well as the relation

$$\frac{\mathrm{Re}_h}{\mathrm{Re}_c} = \frac{G_h \mu_c d_{eq.in} F_c}{G_c \mu_h d_{eq.out} F_h} \tag{8.66}$$

we may obtain

$$N_\alpha = \frac{\alpha_h}{\alpha_c} = \frac{(\mathrm{Nu}_h/\mathrm{Nu}_{h.sm})_{\mathrm{Re}_h}}{(\mathrm{Nu}_c/\mathrm{Nu}_{c.sm})_{\mathrm{Re}_c}} \frac{S_c}{S_h} \left(\frac{d_{in}}{d_{out}}\right)^{0.2} \left(\frac{T_w}{T_c}\right)^{0.55} \left(\frac{G_h \mu_c}{G_c \mu_h}\right)^{0.8} \frac{\lambda_h}{\lambda_c} \tag{8.67}$$

Here, S_c and S_h are the areas of flow cross sections on the cold and hot sides of the heat exchanger; d_{in} and d_{out} are the maximum sizes of the oval on the inside and outside of the helical tube. The ratios $(\mathrm{Nu}_{sm}/\mathrm{Nu}_{h.sm})_{\mathrm{Re}_h}$ and $(\mathrm{Nu}_c/\mathrm{Nu}_{c.sm})_{\mathrm{Re}_c}$ characterize an increase in the heat transfer coefficient on the inside and on the outside of the tubes, as compared to a smooth tube. The subscripts "Re" in these ratios denote that these are taken at the appropriate Re for helical tubes.

Analysis of the above expressions shows that since the numbers Re_c and Re_h are related by Eq. (8.66), a heat exchanger can be calculated for one of the hydraulic resistances Δp_h and Δp_c (or for their sum). Then, Re_c is calculated by Eq. (8.63) if Δp_c is prescribed, or Re_h is calculated by expression (8.49) if p_h is specified. Then, Re_h or Re_c may be found by Eq. (8.66). The areas of flow cross sections can be determined by

$$S_h = G_h d_{eq.in}/\mathrm{Re}_h \mu_h \tag{8.68}$$

$$S_c = G_c\, d_{eq.out}/\mathrm{Re}_c \mu_c \tag{8.69}$$

The number of heat exchanger tubes N is found by Eq. (8.52), the heating surface F_h by (8.54), and the tube length l by (8.53).

The value of N_α necessary to determine Re_c or Re_h is found from Eq. (8.67). In this case, S_c/S_h is specified by the relationship between the maximum size of the oval and the height of the helical tube cross section.

Thus, the proposed method allows the determination of the sizes of a longitudinal-flow tube bundle heat exchanger with specified pressure losses.

The crossflow tube bundle heat exchanger. Let us consider using as an example a heat exchanger which is multi-pass in the hot fluid inside the tube and is single-pass in the cold one on the outside of the tubes (Fig. 8.2).

The initial data are the inlet hot fluid temperature T'_h, outlet hot fluid temperature T''_h, its flowrate G_h, inlet pressure p'_h, hydraulic resistance Δp_h, inlet temperature T'_c of the cold fluid, outlet temperature T''_c of the cold fluid, its flowrate G_c, inlet pressure p'_c and hydraulic resistance Δp_c.

The geometrical parameters of the helical tubes are assigned. These tubes form a tube bundle: $d_{eq.in}$ is the equivalent inner tube diameter, Π_{in} and Π_{out} are the inner and outer tube perimeters, f_{in} is the area of the flow cross section inside the tubes, f_{area} is the cross-sectional area of a helical tube (determined around the outer perimeter), d_{out} and d_{in} are the maximum outer and inner sizes of the oval, s_2 and s_1 are the longitudinal and transverse tube bundle pitches, ψ is the tube bundle porosity with respect to the fluid determined by (7.3), and d_s is the characteristic dimension of the intertube space determined by (7.4).

The exchanger heat power Q, mean logarithmic temperature difference ΔT_{log}, mean temperatures $\bar{T}_h$ and $\bar{T}_c$ of the fluids, and the mean hot and cold fluid densities $\bar{\rho}_h$ and $\bar{\rho}_c$ are determined.

First, let us consider the heat transfer and hydraulic resistance in the cold heat carrier flowing in the intertube space. As discussed in Chapter 7, the mean heat transfer coefficient for helical tube bundles in crossflow may be represented as:

$$\alpha_c = \frac{C_1 \lambda_c \mathrm{Re}_c^n}{d_s} = K_c\, \mathrm{Re}_c^n \tag{8.70}$$

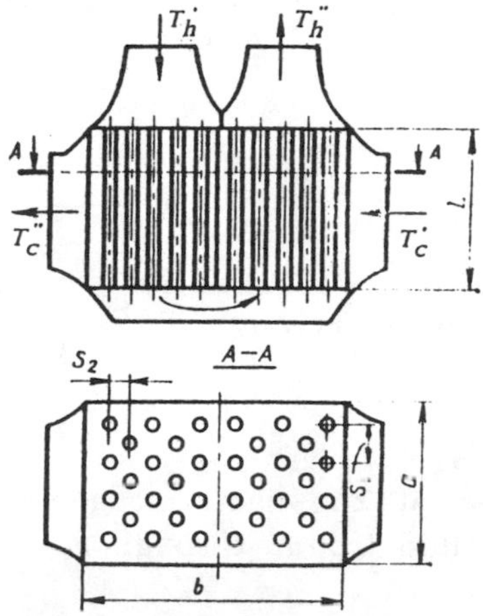

Figure 8.2 Schematic of a two-pass heat exchanger

where λ_c is the thermal conductivity for the cold heat carrier at a temperature $\bar{T}_c$; C_1 is a constant varying within 0.232–0.74; $\mathrm{Re}_c = \bar{\rho}_c \bar{u}_c d_з/\mu_c$ is the Reynolds number; $\bar{u} = \bar{u}_{0c}/\psi$ is the characteristic velocity; $\bar{u}_0$ is the velocity ahead of a tube bundle; μ_c is the dynamic viscosity at the temperature $\bar{T}_c$; and K_c is a constant. The exponent of Re_c varies as n = 0.58–0.62 with Re_c ranging from 10^3 to $3{\cdot}10^4$.

The hydraulic resistance of helical tube bundles in crossflow may be expressed as:

$$\Delta p_c = \mathrm{Eu}_c \bar{\rho}_c \bar{u}_c^2 z = C_2 \mathrm{Re}_c^m \bar{\rho}_c \bar{u}_c^2 z_c \tag{8.71}$$

where C_2 is a constant varying within 1.18–8 and z is the number of tube rows.

The exponent m varies from 0–0.156 at Re_c = 10^3–$3{\cdot}10^4$. Let us determine the pressure losses in a tube bundle in terms of the conventional resistance coefficient

$$\xi_{\mathrm{res.c}} = \frac{\Delta p_c}{\bar{\rho}_c \bar{u}_c^2/2} = 2\mathrm{Eu}_c z = 2C_2 \mathrm{Re}_c^m z \tag{8.72}$$

The number of tube rows z is unknown. The following relations are used for its determination. Let us designate the fluid amounts by a, b, and l (Fig. 8.2). The heating surface determined with respect to the outer tube perimeter will be equal to:

$$F_{\text{out}} = \frac{abl\,\Pi_{\text{out}}}{s_1 s_2} \tag{8.73}$$

and the conventional area of the flow cross section of the intertube space determined through $\bar{u}_c$ is

$$S_c = al\psi \tag{8.74}$$

and the frontal area of the heat exchanger determined with respect to the cold fluid (found through $\bar{u}_c$) is

$$S_{0c} = al \tag{8.75}$$

Hence, it is possible to determine the heat exchanger length along the gas flow in the intertube space

$$B = \frac{F_{\text{out}}\,s_1 s_2}{S_{0c}\,\Pi_{\text{out}}} \tag{8.76}$$

but since

$$S_{0c} = \frac{Q}{c_{pc}(T''_c - T'_c)\,\psi\bar{u}_c\bar{\rho}_c} \tag{8.77}$$

$$F_{\text{out}} = Q/\alpha_c \Delta T_c \tag{8.78}$$

then

$$b = \frac{\bar{u}_c\bar{\rho}_c c_{pc}\,s_1 s_2\,\psi}{\alpha_c\,\Pi_{\text{out}}}\;\frac{T''_c - T'_c}{\Delta T_c} \tag{8.79}$$

Here, ΔT_c is the mean temperature difference on the cold side, and c_{pc} is the mean heat capacity of cold fluid.

Considering the relation

$$\text{Re}_c = \frac{\bar{\rho}_c\bar{u}_c d_s}{u_c} = \frac{\bar{\rho}_c\bar{u}_{0c}\cdot 2f_{\text{area}}}{\mu_c(1-\psi)\,s_2} \tag{8.80}$$

we arrive at:

$$b = \frac{\mathrm{Re}_c \mu_c c_{pc} (1-\psi) s_1 s_2^2}{\alpha_c \cdot 2\Pi_{out} f_{area}} \; \frac{T''_c - T'_c}{\Delta T_c} \tag{8.81}$$

Determining α_c by Eq. (8.70) and using $\mathrm{Pr}_c = \mu_c c_{pc}/\lambda_c$, we have:

$$b = \frac{\mathrm{Re}_c^{1-n} \mathrm{Pr}_c S_1 S_2 \psi}{C_1 \Pi_{out}} \; \frac{T''_c - T'_c}{\Delta T_c} \tag{8.82}$$

Since

$$z = \frac{b}{S_2} = \frac{\mathrm{Re}_c^{1-n} \mathrm{Pr}_c S_1 \psi}{C_1 \Pi_{out}} \; \frac{T''_c - T'_c}{\Delta T_c} \tag{8.83}$$

then

$$\xi_{res.c} = \frac{2 C_2 \mathrm{Re}_c^{1-n+m} \mathrm{Pr}_c S_1 \psi}{C_1 \Pi_{out}} \; \frac{T''_c - T'_c}{\Delta T_c} \tag{8.84}$$

In actual heat exchangers, the number Re_c usually varies from $5 \cdot 10^3$ to 10^4. Within these ranges of Re_c for the tube bundles considered, m = 0.13–0.17, n = 0.62–0.58, and $1-n+m$ = 0.25. Re_c^{1-n+m} varies from 8.3 to 10. Let us assume for further calculations that $\mathrm{Re}_c^{1-n+m} = 9.15$, then we have

$$\xi_{res.c} = \frac{18.3 C_2 \mathrm{Pr}_c S_1 \psi}{C_1 \Pi_{out}} \; \frac{T''_c - T'_c}{\Delta T_c} = \frac{K_1}{\Delta T_c} \tag{8.85}$$

Neglecting the temperature difference in the tube wall gives:

$$\Delta T_c = \frac{N_\alpha}{N_\alpha + d_{out}/d_{in}} \, \Delta \bar{T}_{log} \tag{8.86}$$

where $N_\alpha = \alpha_h/\alpha_c$ is the relationship between the heat transfer coefficients on the hot and cold sides. A knowledge of $\xi_{res.c}$ enables one to determine the Reynolds number for the intertube space:

$$\mathrm{Re}_c^2 = \frac{4(1-\sigma_c^2)\bar{\rho}_c^2 d_c^2 R_c \bar{T}_c}{(1+\sigma_c)2\mu_c^2\left[\xi_{\mathrm{res.c}} + \dfrac{2(T_c'' - T_c')}{\bar{T}_c} + 4\dfrac{1-\sigma_c}{1+\sigma_c}\right]} \tag{8.87}$$

where $\sigma_c = 1-\Delta p_c/p_c'$, and R_c is the gas constant.

Now, let us examine the tube flow. The pressure losses in a gas flowing in a tube include friction losses, local resistances, and flow accelerations. When the varying gas velocity effect on the local resistances is neglected, the pressure losses may be expressed as:

$$\Delta p_h = \left[\xi_h \frac{l}{d_{eq.in}} m_h + \Sigma\xi_i + 2\frac{T_h'' - T_h'}{\bar{T}_h} + 4\frac{1-\sigma_h}{1+\sigma_h}\right]\frac{\bar{\rho}_h \bar{u}_h^2}{2} \tag{8.88}$$

where ξ_h is the hydraulic resistance coefficient, l the tube length, m_h the number of passes, ξ_i the local resistance coefficient, $\bar{u}_h$ the tube mean gas velocity, and $\sigma_h = 1-\Delta p_h/p_h'$.

In [36], Eq. (8.88) was reduced to the form:

$$\sigma_h = \sqrt{1 - \frac{\mathrm{Re}_h^2(1+\sigma_h)\mu_h^2}{4\bar{\rho}_h^2 d_{\mathrm{eq.in}}^2 R_h \bar{T}_h}\left[\xi_h \frac{l}{d_{\mathrm{eq.in}}} m_h + \Sigma\xi_i + \frac{2(T_h'' - T_h')}{\bar{T}_h} + 4\frac{1-\sigma_h}{1+\sigma_h}\right]} \tag{8.89}$$

where μ_h is the dynamic viscosity at a temperature $\bar{T}_h$, Re_h is the Reynolds number on the hot side, and R_h is the gas constant.

From Eq. (8.89) we may obtain:

$$\mathrm{Re}_h^2 = \frac{4(1-\sigma_h^2)\bar{\rho}_h^2 d_{\mathrm{eq.in}}^2 R_h \bar{T}_h}{(1+\sigma_h)^2\mu_h^2\left[\xi_h \dfrac{l}{d_{\mathrm{eq.in}}} m_h + \Sigma\xi_i + \dfrac{2(T_h'' - T_h')}{\bar{T}_h} + 4\dfrac{1-\sigma_h}{1+\sigma_h}\right]} \tag{8.90}$$

In Eq. (8.90), the product $\xi_h\ (l/d_{\mathrm{eq.in}})m_h$ is unknown and depends weakly on the Reynolds number. For helical tubes, with regard to heat transfer enhancement, we have:

$$\xi_h = 0.3164/\mathrm{Re}_h^{0.25}\left(\frac{\xi_h}{\xi_{h.sm}}\right)_{\mathrm{Re}_h} \tag{8.91}$$

Equation (8.91) is valid for the turbulent regime both for isothermal flow and for gas cooling.

Let us designate the area of the flow cross section of one pass by S_h, the total number of tubes by N, and the heating surface determined with respect to the inner tube perimeter by F_h. Since

$$l = F_h/(\Pi_{in} N) \tag{8.92}$$

$$N = 4S_h\, m_h/(\Pi_{in}\, d_{eq.in}) \tag{8.93}$$

it may be stated that

$$l/d_{eq.in} = F_h/(4S_h m_h) \tag{8.94}$$

The heat flux transferred through the heating surface may be represented as:

$$Q = \alpha_h F_h \Delta T_h \tag{8.95}$$

where ΔT_h is the mean temperature difference on the hot side.

On the other hand,

$$Q = G_h c_h (T'_h - T''_h) = \bar{u}_h \bar{\rho}_h S_h c_h (T'_h - T''_h) \tag{8.96}$$

where c_{ph} is the mean heat capacity. Using Eqs. (8.94)–(8.96), we arrive at:

$$\frac{l}{d_{eq.h}} m_h = \frac{\bar{u}_h \bar{\rho}_h c_h (T'_h - T''_h)}{4\alpha_h \Delta T_h} = \frac{\mathrm{Re}_h \mu_h c_h (T'_h - T''_h)}{4 d_{eq.in} \alpha_h \Delta T_h} \tag{8.97}$$

For turbulent gas flow in a helical tube and gas cooling, the heat transfer coefficient may be expressed as:

$$\alpha_h = \left(\frac{\mathrm{Nu}_h}{\mathrm{Nu}_{h.sm}}\right)_{\mathrm{Re}_h} \frac{0.023\, \mathrm{Re}_h^{0.8} \mathrm{Pr}_h^{0.4} \lambda_h}{d_{eq.in}} = K_h \mathrm{Re}_h^{0.8} \tag{8.98}$$

where λ_h is the thermal conductivity.

Substituting Eq. (8.98) into (8.97) and using the definition of the Prandtl number $\mathrm{Pr}_h = \mu_h c_{ph}/\lambda_h$ results in:

$$\frac{l}{d_{eq.in}} m_h = \frac{\mathrm{Re}_h^{0.2} \mathrm{Pr}_h^{0.6} (T'_h - T''_h)}{0.092 \Delta T_h (\mathrm{Nu}_h/\mathrm{Nu}_{h.sm})_{\mathrm{Re}_h}} \tag{8.99}$$

or, using (8.91), we have

$$\xi_h \frac{l}{d_{eq.in}} m_h = \frac{3.44\,\mathrm{Pr}_h^{0.6}\,(T'_h - T''_h)}{\mathrm{Re}_h^{0.05}\,\Delta T_h}\;\frac{(\xi_h/\xi_{h.sm})\mathrm{Re}_h}{(\mathrm{Nu}_h/\mathrm{Nu}_{h.sm})\mathrm{Re}_h}$$

$$= \frac{1.75\,(T'_h - T''_h)}{\Delta T_h}\;\frac{(\xi_h/\xi_{h.sm})\mathrm{Re}_h}{(\mathrm{Nu}_h/\mathrm{Nu}_{h.sm})\mathrm{Re}_h} = \frac{K_2}{\Delta T_h} \tag{8.100}$$

since for a gas $\mathrm{Pr}_h = 0.71$, and for $\mathrm{Re}_h = 10^4\text{–}10^5$, $\mathrm{Re}_h^{0.05} = 1.6$ is taken. Using (8.86) we may obtain

$$\Delta T_h = \frac{d_{out}/d_{in}}{N_\alpha + d_{out}/d_{in}}\;\Delta \bar{T}_{log} \tag{8.101}$$

Applying Eqs. (8.70) and (8.98) to the heat transfer coefficients, Eqs. (8.87) and (8.90) for the Reynolds number, as well as Eqs. (8.86), (8.100), (8.101), and (8.85), we arrive at the equation for the determination of an unknown quantity N_α:

$$\frac{1}{N_\alpha} = \frac{K_c\left\{\dfrac{4(1-\sigma_c^2)\bar{\rho}_c^2 d_c^2 R_c \bar{T}_c}{(1+\sigma_c)^2\mu_c^2\left[\dfrac{K_1}{\Delta\bar{T}_{log}N_\alpha}\left(N_\alpha + \dfrac{d_{out}}{d_{in}}\right) + 2\dfrac{T''_c - T'_c}{\bar{T}_c} + 4\dfrac{1-\sigma_c}{1+\sigma_c}\right]}\right\}^{n/2}}{K_h\left\{\dfrac{4(1-\sigma_h^2)\bar{\rho}_h^2 d_{eq.in}^2 R_h \bar{T}_h}{(1+\sigma_h)^2\mu_h^2\left[\dfrac{K_2 d_{in}}{\Delta\bar{T}_{log}d_{out}}\left(N_\alpha + \dfrac{d_{out}}{d_{in}}\right) + \Sigma\xi_i + 2\dfrac{T''_h - T'_h}{\bar{T}_h} + 4\dfrac{1-\sigma_h}{1+\sigma_h}\right]}\right\}^{0.4}} \tag{8.102}$$

Equation (8.102) can be solved simply by a graphical technique. A knowledge of N_α enables one first to determine Re_c, Re_h, and then the areas of the flow cross sections in the intertube space

$$S_c = (G_c d_s)/(\mathrm{Re}_c \mu_c)$$

$$S_{0c} = S_c/\psi \tag{8.103}$$

and inside the tubes

$$S_h = (G_h\, d_{eq.in})/(\mathrm{Re}_h \mu_h) \tag{8.104}$$

Then, employing Eq. (8.93) it is possible to find the number of heat exchanger tubes; using (8.70) and (8.98) it is possible to find the heat transfer coefficients α_c and α_h; and using the heat transfer coefficients, the heating surface F_{in} can be determined with respect to the inner tube perimeter. Using the relations $N = ab/s_1 s_2$

$$F_{out} = (abl\Pi_{out})/(s_1 s_2), \; s_c = al\psi$$

we may determine the sizes of the heat exchanger

$$l = F_{out}/(\Pi_{out} N) \tag{8.105}$$

$$a = S_c/(l\psi) = S_{0c}/l \tag{8.106}$$

$$b = Ns_1 s_2/a \tag{8.107}$$

The number of tube rows in the gas direction is

$$z = b/s_2 = Ns_1/a \tag{8.108}$$

Thus, the above method allows direct determination of a crossflow tube bundle heat exchanger at assigned pressure losses in the intertube space and inside the tubes.

The above method was derived for the most widely encountered case, namely, one-pass flow in the intertube space. This method can easily be extended to multipass flow in the intertube space. For this, Eqs. (8.71) and (8.72) are supplemented with a factor m_c equal to the number of passes, and will be equal to the number of tube rows per one pass, i.e.,

$$\xi_{res.c} = 2C_2 \mathrm{Re}_c^m z m_c \tag{8.109}$$

The frontal area of the heat exchanger with respect to the cold fluid will be:

$$S_{0c} = al/m_c \tag{8.110}$$

and the length of one pass will be:

$$b = \frac{\bar{u}_c \bar{\rho}_c c_{pc} s_1 s_2 \psi}{\alpha_c \Pi_{out} m_c} \; \frac{T''_c - T'_c}{\Delta T_c} \tag{8.111}$$

and the number of tube rows per one pass will be:

$$z = \frac{b}{S_2} = \frac{\mathrm{Re}_c^{1-n}\,\mathrm{Pr}_c\, s_1 \psi}{C_1 \Pi_{out} m_c} \; \frac{T_c'' - T_c'}{\Delta T_c} \tag{8.112}$$

Equation (8.84) for $\xi_{res.c}$ remains, therefore, invariable.

Equation (8.87) for Re_c must be supplemented with a sum of the local resistance coefficients $\Sigma\xi_i$ at the turn of the flow between the passes

$$\mathrm{Re}_c^2 = \frac{4(1-\sigma_c^2)\bar{\rho}_c^2 d_s^2 R_c \bar{T}_c}{(1+\sigma_c)^2 \mu_c^2 \left[\xi_{res.c} + \Sigma\xi_i + \dfrac{2(T_c'' - T_c')}{\bar{T}_c} + 4\dfrac{1-\sigma_c}{1+\sigma_c}\right]} \tag{8.113}$$

Further calculations are the same, but Eq. (8.106) must be replaced by:

$$a = m_c S_c / l\psi = m_c S_{0c} / l \tag{8.114}$$

8.4 ESTIMATION OF THE EFFICIENCY OF HELICAL TUBES USED IN HEAT EXCHANGERS

The relations considered in section 8.1 can be used to estimate the efficiency of helical tubes. A comparison of the volumes of helical and smooth tube heat exchangers is made at equal tube perimeters, heat powers, fluid flowrates, and hydraulic resistance.

First, let us consider a longitudinal-flow tube bundle heat exchanger. If in such a heat exchanger the thermal resistance inside the tubes (heat transfer coefficient inside the tubes is considerably less than in the intertube space) is limiting, then V/V_{sm} is determined using the data of Chapter 6 for the flow inside helical tubes. At a tube twisting pitch $s/d = 6.2$ we have $\mathrm{Nu}/\mathrm{Nu}_{sm} = 1.4$ and $\xi/\xi_{sm} = 1.7$, which, according to Eq. (8.14), results in $V/V_{sm} = 0.8$. The heat exchanger efficiency decreases with increasing tube twisting pitch. At $s/d = 16.7$ we have $\mathrm{Nu}/\mathrm{Nu}_{sm} = 1.2$ and $\xi/\xi_{sm} = 1.30$, $V/V_{sm} = 0.89$.

The helical tubes prove to be more effective if the thermal resistance in the intertube space is limiting. Figure 8.3 plots $\mathrm{Nu}/\mathrm{Nu}_{sm}$ and ξ/ξ_{sm} for a helical tube bundle in longitudinal flow at $\mathrm{Re} > 10^4$. At $s/d > 12$, $\mathrm{Nu}/\mathrm{Nu}_{sm} \approx \xi/\xi_{sm}$, and at smaller values of s/d the increase in ξ/ξ_{sm} is greater

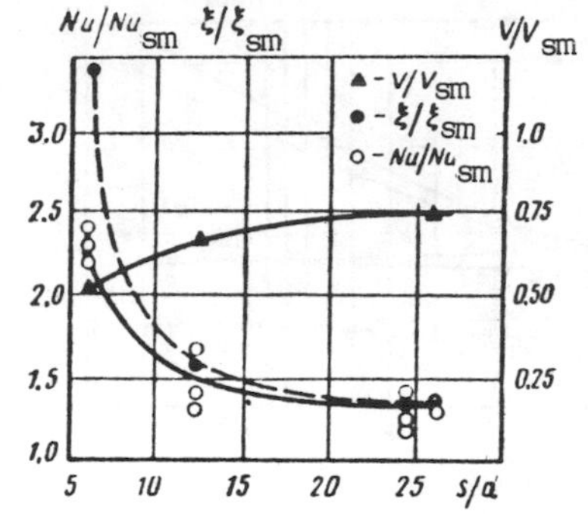

Figure 8.3 Nu/Nu_{sm}(○), ξ/ξ_{sm}(●), and V/V_{sm}(▲) versus s/d for a longitudinal-flow helical tube heat exchanger with Re > 10^4

than the increase of Nu/Nu_{sm}. At $s/d = 6$, $Nu/Nu_{sm} \approx 2.2$ for $\xi/\xi_{sm} \approx 3.4$. This figure also shows V/V_{sm} versus s/d. As seen from the figure, V/V_{sm} varies from 0.52 to 0.74 and decreases with decreasing s/d. At smaller values of Re, the efficiency of helical tubes will be still higher. For example, at Re = $3 \cdot 10^3$ and $s/d = 12$, $Nu/Nu_{sm} = 1.75$ and $\xi/\xi_{sm} = 1.5$. At the same Re and $s/d = 6$, $Nu/Nu_{sm} \approx 3.5$ at $\xi/\xi_{sm} \approx 3.8$, which allows a 2–2.5-fold decrease in the heat exchanger volume. For an arbitrary relationship between the coefficients for heat transfer on the outside and on the inside of the tubes, the ratio V/V_{sm} is determined by Eq. (8.22).

The above calculations have shown that $s/d = 6$ (from the considered pitch range of s/d) is an optimum tube twisting pitch. At smaller s/d, a sharp increase in the hydraulic resistance will augment V/V_{sm}.

The relations cited in section 8.2 can be used to calculate the efficiency of a crossflow tube bundle heat exchanger. Let us consider a bundle in which the tubes are arranged in a transverse row with approximately constant gaps between them. These gaps form, along the tube bundle, slotted channels with a maximum width equal to half the difference between the minimum and maximum sizes of the oval. The tubes contact only those of a neighboring row. Figure 8.4 plots Nu/Nu_{sm} and ξ/ξ_{sm} versus Re for such a tube bundle. The data for smooth tubes are taken from [19]. As seen from this figure, the replacement of smooth tubes by helical ones allows a 1.3–1.45-fold increase in heat transfer when the hydraulic resistance increases 1.4–2.5 times, which makes it possible to reduce the volume of a heat exchanger by 7–27%.

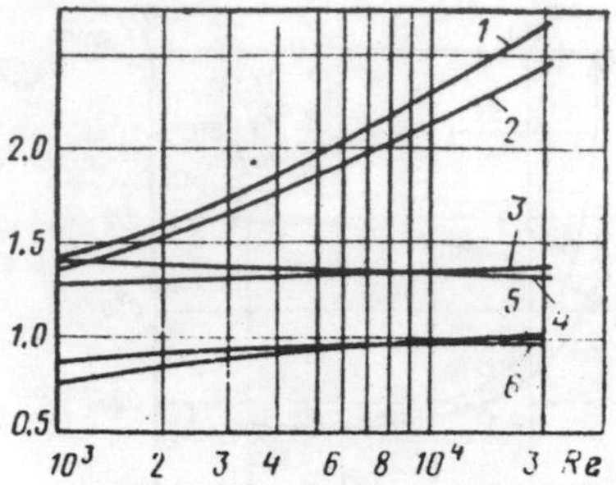

Figure 8.4 ξ/ξ_{sm} (*1, 2*), Nu/Nu_{sm} (*3, 4*), V/V_{sm} (*5, 6*) for a crossflow helical tube heat exchanger: *1, 3, 5*) s/d = 6.1; *2, 4, 6*) s/d = 1.2

The efficiency of the helical tubes decreases with increasing Re.

Thus, the above-mentioned estimates show that the use of helical tubes enables one to substantially reduce the volume of cross- and longitudinal-flow tube bundle heat exchangers. The efficiency is maximum for those heat exchangers in which the heat transfer coefficient for the intertube space is a minimum.

REFERENCES

1. Abramovich, G. N. Teoriya turbulentynykh struy (The theory of turbulent jets). Gosizdat. Fiz.-Mat. Lit., Moscow, 1960, 715 pp.
2. Ashmantas, L. A., Vilemas, Yu. V., and Dzyubenko, B. V. Intensifikatsiya teploobmena v poperechno obtekayemykh puchkakh vitykh trub (Heat transfer enhancement in crossflow past helical tube bundles). Izv. AN SSSR, Energetika i Transport, 1982, No. 6, pp. 120-127.
3. Bobkov, V. P., Ibragimov, M. Kh., and Sabelev, G. I. Obobshcheniye eksperimental'nykh dannykh po intensivnosti pul'satsiy skorosti pri turbulentnom techenii zhidkosti v kanalakh razlichnoy formy (Correlation of experimental data on the intensity of velocity pulsations in a turbulent flow in different-geometry channels). Izv. AN SSSR, Ser. Mikh. Zhidkosti i Gaza, 1968, No. 3, pp. 162-173.
4. Borishanskiy, V. M., Gotovskiy, M. A., and Mizonov, N. V. Metod gomogennovo potoka i yevo primeneniye dlya rascheta gidrodinamiki i teploperedachi v puchkakh sterzhney. V: Teploobmen i gidrodinamika odnofaznovo potoka v puchkakh sterzhney (The homogeneous flow method as applied to the calculation of flow dynamics and heat transfer in rod bundles. In: Heat transfer and flow dynamics of single-phase flow in rod bundles). Nauka, Leningrad, 1979, pp. 22-41.
5. Varshkyavicius, R. R. Gidravlicheskoye soprotivleniye i osobennosti techeniya v mezhtrubnom prostranstve teploobmennika s zakrutkoy vitykh trub oval'novo profilya - Mezhvuzovskiy sbornik nauchnykh trudov: Sovremennyye problemy gidrodinamiki i teploobmena v elementakh

energeticheskikh ustanovok i kriogennoy tekhnike (Hydraulic resistance and specific features of the flow in the intertube space of a heat exchanger with twisted oval-shaped tube. Collected papers of the All-Union Correspondence Mechanical Engineering Institute: Modern problems of flow dynamics and heat transfer in the elements of power plants and in cryogenic techniques). All-Union Correspondence Mechanical Engineering Institute, Moscow, 1983, No. 12, pp. 100-106.
6. Varshkyavičius, R. R., Vilemas, Yu. V., and Poshkas, P. S. Gidravlicheskoye soprotivleniye puchkov profilirovannykh vintoobraznykh trub v poperechnom potoke gaza (Hydraulic resistance of profiled helical tube bundles in crossflow). Trudy AN Lit. SSR, 1977, Ser. B, Vol. 6, No. 103, pp. 55-61.
7. Vilemas, Yu. V., Chesna, B. A., and Survila, V. Yu. Teplootdacha v gazookhlazhdayemykh kol'tsevykh kanalakh (Heat transfer in gas-cooled circular channels). Mokslas, Vilnius, 1977, 253 pp.
8. Vilemas, Yu. V., Dzyubenko, B. V., and Sakalauskas, A. V. Issledovaniye struktury potoka v teploobmennikakh s vintoobrazno zakruchennymi trubami (Study of the flow structure in heat exchangers with helically twisted tubes). Izv. AN SSSR, Energetika i Transport, 1980, No. 4, pp. 135-144.
9. Vilemas, Yu. V., Dzyubenko, B. V., and Sakalauskas, A. V. Struktura potoka i yevo perenosnyye svoystva v teploobmennike s vintoobrazno zakruchennymi trubami - Mezhvuzovskiy sbornik nauchnykh trudov: Sovremennyye problemy gidrodinamiki i teploobmena v elementakh energeticheskikh ustanovok k kriogennoy tekhnike (The structure of the flow and its transport properties in a helical tube bundle. Collected papers of the All-Union Correspondence Mechanical Engineering Institute: Modern problems of flow dynamics and heat transfer in the elements of power plants and in cryogenic technique). All-Union Correspondence Mechanical Engineering Institute, Moscow, 1981, No. 10, pp. 3-14.
10. Vilemas, Yu. V., Shimonis, V. M., and Nyamira, M. A. Teploobmen v kol'tsevom kanale s prepyatstviyem pri turbulentom techenii vozdukha s peremennymi fizicheskimi svoystvami (Heat transfer in a circular channel with an obstacle for turbulent flow of air with variable physical properties). Trudy AN Lit. SSR, 1979, Ser. B, Vol. 2 (111), pp. 81-87.

11. Gershgorin, S. A. O priblizhennom integrirovanii differentsial'nykh uravneniy Laplasa i Puassono (On the approximate integration of the differential Laplace and Poisson equations). Izv. Leningrad Polytekh. Inst., 1927, Vol. 30, p. 75.
12. Dzyubenko, B. V., Vilemas, Yu. V., and Ashmantas, L. A. Peremeshivaniye teplonositelya v teploobmennike s zakrutkoy potoka (Heat carrier mixing in a heat exchanger with flow swirling). J. Eng. Physics (Russian), 1981, Vol. 40, No. 5, pp. 773-779.
13. Dzyubenko, B. V. and Dreytser, G. A. Issledovaniye teploobmena i gidravlicheskovo soprotivleniya v teploobmennom apparate s zakrutkoy potoka (Study of heat transfer and hydraulic resistance in a heat exchanger with flow swirling). Izv. AN SSR, Energetika i Transport, 1979, No. 5, pp. 163-171.
14. Dzyubenko, B. V. and Ievlev, V. M. Teploobmen i gidravlicheskoye soprotivleniye v mezhtrubnom prostranstve teploobmennika s zakrutkoy potoka (Heat transfer and hydraulic resistance in the intertube space of a heat exchanger with flow swirling). Izv. AN SSSR, Energetika i Transport, 1980, No. 5, pp. 117-125.
15. Dzyubenko, B. V., Sakalauskas, A. V., and Vilemas, Yu. V. Raspredeleniya skorostey i staticheskovo davleniya v teploobmennike s zakrutkoy potoka (Velocity and static pressure distributions in a heat exchanger with swirling flow). Izv. AN SSSR, Energetika i Transport, 1981, No. 4, pp. 112-118.
16. Dzyubenko, B. V., Sakalauskas, A. V., and Vilemas, Yu. V. Energeticheskiye spektry turbulentnosti v teploobmennike s zakrutkoy potoka (Energy turbulence spectra in a heat exchanger with swirling flow). Izv. AN SSSR, Energetika i Transport, 1983, No. 4, pp. 125-133.
17. Dzyubenko, B. V., Sakalauskas, A. V., and Vilemas, Yu. V. Mestnaya teplootdacha v mezhtrubnom prostranstve teploobmennovo apparata s zakrutkoy potoka (Local heat transfer in the intertube space of a heat exchanger with swirling flow). J. Eng. Phys. (Russian), 1981, Vol. 41, No. 2, pp. 197-202.
18. Dzyubenko, B. V., Urbonas, P. A., and Ashmantas, L. A. Mezhkanal'noye peremeshivaniye teplonositelya v puchke vitykh trub (Interchannel mixing of a heat carrier in a helical tube bundle). J. Eng. Phys. (Russian), 1983, Vol. 45, No. 1, pp. 26-32.

19. Žukauskas, A. A. Konvektivnyy perenos v teploobmennikakh (Convective transfer in heat exchangers). Nauka, Moscow, 1982, 427 pp.
20. Žukauskas, A. A., Makaryavičius, V. I., and Shlanchyauskas, A. A. Teplootdacha puchkov trub v poperechnom potoke zhidkosti (Heat transfer in tube bundles with liquid crossflow). Mintis, Vilnius, 1968, 192 pp.
21. Zdanavičius, G. B., Chesna, B. A., and Žyugzhda, I. I. Mestnaya teplootdacha poperechno obtekayemovo potokom vozdukha kruglovo tsilindra pri bol'shikh znacheniyakh Re (Local heat transfer from a circular cylinder in air crossflow at large Re). Trudy Lit. SSR, Ser. B, 1975, Vol. 2, No. 87, pp. 109–119.
22. Idel'chik, I. E. Spravochnik po gidravlicheskim soprotivleniyam (Handbook of hydraulic resistance). Mashinostroyeniye, Moscow, 1975, 558 pp.
23. Ievlev, V. M. Turbulentnoye dvizheniye vysokotemperaturnykh sploshnykh sred (Turbulent motion of high-temperature continua). Nauka, Moscow, 1975, 256 pp.
24. Ievlev, V. M., Danilov, Yu. I., and Dzyubenko, B. V. Teploobmen i gidrodinamika zakruchennykh potokov v kanalakh slozhnoy formy (Heat transfer and flow dynamics of swirled flows in complex-geometry channels). V: Teplomassoobmen-VI (In: Heat and mass transfer VI). Heat and Mass Transfer Institute, 1980, Vol. 1, Pt. 1, pp. 88–99.
25. Ievlev, V. M., Dzyubenko, V. B., and Segal', M. D. Teplomassoobmen v teploobmennike s zakrutkoy potoka (Heat and mass transfer in a heat exchanger with swirling flow). Izv. AN SSSR, Energetika i Transport, 1981, No. 5, pp. 104–112.
26. Kalinin, E. K., Dreytser, G. A., and Yarkho, S. A. Intensifikatsiya teploobmena v kanalakh (Heat transfer improvement in channels). Mashinostroyeniye, Moscow, 1981, 205 pp.
27. Kast, W., Krischer, O., and Reinicke, H. Konvektive Wärme- und Stoffubergang. Springer-Verlag, Berlin, 1974.
28. Kidd, G. I. Teplootdacha k gazovomy potoku i padeniye davleniya v spiral'no-volnistykh trubakh (The heat transfer and pressure-drop characteristics of gas flows inside spirally corrugated tubes). Trans. ASME, Journal of Heat Transfer, Ser. C, 1970, Vol. 92, No. 3, pp. 205–211.

29. Conte-Bellot, G. Ecoulement turbulent entre deux parois paralleles. Publications Scientifiques et Techniques du Ministere de L'Air, Paris, 1965.
30. Koshkin, V. K. and Kalinin, E. K. Teploobmennyye apparaty i teplonositeli (Heat exchangers and heat carriers). Mashinostroyeniye, Moscow, 1971, 200 pp.
31. Koshmarov, Yu. A. Gidrodinamika i teploobmen turbulentnovo potoka neszhimayemoy zhidkosti v zazore mezhdu vrashchayushchimisya koaksial'nymi tsilindrami (Flow dynamics and heat transfer of turbulent incompressible liquid flow in the gap between rotating coaxial cylinders). J. Eng. Phys. (Russian), 1962, Vol. 5, No. 5, pp. 5-14.
32. Kutateladze, S. S. Osnovy teorii teploobmena (Fundamentals of heat transfer theory). Atomizdat, Moscow, 1979, 416 pp.
33. Kutateladze, S. S. and Borishanskiy, V. M. Spravochnik po teploperedache (Handbook on heat transfer). Gosenergoizdat, Moscow, 1959, 414 pp.
34. Migay, V. K. Gidravlicheskoye soprotivleniye treugol'nykh kanalov v laminarnom potoke (Hydraulic resistance of triangular channels in longitudinal flow). Izv. VUZov, Ser. Energetika, 1963, No. 5, pp. 122-124.
35. Minskiy, E. M. and Fomichev, M. S. O puti smesheniya Prandtlya, kriteriyakh i masshtabakh turbulentnosti (On the Prandtl mixing length, turbulence numbers and scales). Izv. AN SSSR, Ser. Energetika i Transport, 1972, No. 6, pp. 119-123.
36. Mikhaylov, A. I., Borisov, V. V., and Kalinin, E. K. Gazoturbinnyye ustanovki zamknutovo tsikla (Gas turbine closed-cycle installations). Izv. AN SSSR, Moscow, 1962, 148 pp.
37. Nyamira, M. A. and Vilemas, Yu. V. Teplootdacha tsilindricheskovo i vintoobraznovo sterzhney v kol'tsevom kanale turbulentnomy potoku vozdukha s peremennymi fizicheskimi svoystvami (Heat transfer from cylindrical and helical rods in an annular channel to turbulent air flow with variable physical properties). Trudy AN Lit. SSR, Ser. B, 1975, Vol. 3 (88), pp. 127-136.
38. Dytnerskii, Yu. I., ed. Osnovnyye protsessy i apparaty khimicheskoy tekhnologii (Basic processes and apparatus of chemical engineering). Khimiya, Moscow, 1983, 272 pp.
39. Pavlov, K. F., Romankov, P. G., and Noskov, A. A. Primary i zadachi po kursy protsessov i apparatov khimicheskoy

tekhnologii (Examples and problems of the processes and apparatus of chemical engineering). Khimiya, Leningrad, 1981, 560 pp.

40. Paramonov, N. V. Issledovaniye intensifikatsii teploobmena v profil'nykh trubakh. Teplo- i massoobmen mezhdu potokami i poverkhnostyami. Tematicheskiy sbornik nauchnykh trudov MAI (Study of heat transfer enhancement in profiled tubes. Collected papers of the Moscow Aviation Institute. Heat and mass transfer between the flows and surfaces). 1980, pp. 62-65.
41. Paegle, K. K., Grislis, V. Ya., and Savel'yev, P. A. Intensifikatsiya teploobmena v trubchatykh teploobmennikakh s vintoobraznoy kanavkoy (Heat transfer improvement in tubular heat exchangers with a screw groove). Izv. Lat. SSR, Ser. Fiz. i Tekhn. Nauk, 1967, No. 1, pp. 123-126.
42. Poshkas, P. S. and Survila, V. Yu. Pul'satsii skorosti v mezhtrubnom prostranstve pri poperechnom obtekanii puchkov trub potokom vozdukha (Velocity pulsations in the intertube space in air crossflow past tube bundles). Trudy AN Lit. SSR, Ser. B, 1981, Vol. 1 (122), pp. 71-80.
43. Saul'yev, V. K. Integrirovaniye uravneniy parabolicheskovo tipa metodom setok (Integration of parabolic-type equations by the network method). Fizmatgiz, Moscow, 1960, 324 pp.
44. Segal', M. D. Raschet temperaturnykh poley v r, φ, z-geometrii v poristykh sredakh s prostranstvennoy neravnomernost'yu teplovydeleniya (Calculation of the temperature fields in r, φ, z-geometry in porous media with spatial nonuniform heat release). Preprint No. 2845, Kurchatov Atomic Power Engineering Institute, 1977, 11 pp.
45. Sedov, L. I. Metody podobiya i razmernosti v mekhanike (Similarity and dimension methods as applied in mechanics). Gosizdat Tekhniko-Teor. Lit., Moscow, 1954, 328 pp.
46. Simuni, L. M. Chislennoye resheniye zadachi o neizotermicheskom dvizhenii vyazkoy zhidkosti v ploskoy trube (Numerical solution of the problem of nonisothermal viscous liquid flow in a flat tube). J. Eng. Phys. (Russian), 1966, Vol. 10, No. 1, p. 86.
47. Smitberg, E. and Lendis, F. Treniye i kharakteristiki teploobmena pri vynuzhdennoy konvektsii v trubakh s zavikhritelyma iz skruchennoy lenty (Friction and forced convection heat transfer characteristics in tubes with

twisted tape swirl generators). J. Heat Transfer, Trans. 1964, ASME, Vol. 86, pp. 39-49.

48. Soo, S. Gidrodinamika mnogofaznykh sistem (Fluid dynamics of multiphase systems). Blaisdell Publishing Company, Massachusetts, 1968.

49. Subbotin, V. I., Ibragimov, M. Kh., and Ushakov, P. A. Gidrodinamika i teploobmen v atomnykh energeticheskikh ustanovkakh (Fluid dynamics and heat transfer in atomic power plants). Atomizdat, Moscow, 1975, 408 pp.

50. Survila, V. Yu., Yankauskas, R. I., and Ashmantas, L. A. Gidravlicheskoye soprotivleniye i teplootdacha puchkov plotnoupakovannykh oval'nykh vitykh trub (Hydraulic resistance and heat transfer of close-packed helical oval-shaped tube bundles). Trudy AN Lit. SSR, Ser. B, 1981, Vol. 4 (125), pp. 57-62.

51. Urbonas, P. A. Eksperimental'noye issledovaniye koeffitsienta gidravlicheskovo soprotivleniya v puchke vitykh trub. Mezhvuzovskiy sbornik nauchnykh trudov "Sovremennyye problemy gidrodinamiki i teploobmena v elementakh energeticheskikh ustanovok v kriogennoy tekhnike." (Experimental study of hydraulic resistance coefficients. Collected papers of the All-Union Correspondence Mechanical Engineering Institute: Modern problems of flow dynamics and heat transfer in the elements of power plants and in cryogenic technique). All-Union Correspondence Mechanical Engineering Institute, Moscow, 1982, No. 11, pp. 78-82.

52. Ustimenko, B. P. Protsessy turbulentnovo perenosa vo vrashchayushchikhsya techeniyakh (Turbulent transfer processes in rotating flows). Nauka, Alma-Ata, 1977, 228 pp.

53. Fastovskiy, V. G. and Petrovskiy, Yu. V. Sovremennyye effektivnyye teploobmenniki (Modern effective heat exchangers). Gosenergoizdat, Moscow, 1962, 256 pp.

54. Hinze, J. O. Turbulence. An introduction to its mechanism and theory. McGraw-Hill Book Company, New York, 1959.

55. Schlichting, H. Grenzschicht-theorie. Verglag Braun, Karlsruhe, 1964.

56. Shchukin, V. K. Teploobmen i gidrodinamika vnutrennikh potokov v polyakh massovykh sil (Heat transfer and flow dynamics of internal flows in the body force fields). Mashinostroyeniye, Moscow, 1980, 240 pp.

57. Shchukin, V. K. and Khalatov, A. A. Teploobmen, massoobmen i gidrodinamika zakruchennykh potokov v

osesimmetrichnykh kanalakh (Heat transfer, mass transfer and flow dynamics of swirled flows in axisymmetric channels). Mashinostroyeniye, Moscow, 1982, 200 pp.

58. Ievlev, V. M., Dzyubenko, B. V., Dreytser, G. A., and Vilemas, Yu. V. In-line and crossflow helical tube heat exchangers. Int. J. Heat and Mass Transfer, 1982, Vol. 25, No. 3, pp. 317-323.
59. Ievlev, V. M., Kalinin, E. K., Danilov, Yu. I., Dzyubenko, B. V., and Dreytser, G. A. Heat transfer in the turbulent swirling flow in a channel of complex shape. Proceedings of the seventh Int. Heat Transfer Conference, München, Hemisphere Publishing Corporation, 1982, Vol. 3, General Papers, pp. 171-176.
60. Kjellstrom, B. Studies of turbulent flow parallel to a rod bundle at triangular array. Report of the Swedish Committee on Atomic Energy, Studsvik, 1974, No. AE-487, p. 109.
61. Trupp, A. G. and Azed, R. S. The structure of turbulent flow in triangular array rod bundles. Nuclear Engineering and Design, 1975, Vol. 32, pp. 47-84.

INDEX